# Legal Protection of European Inland Waters

This book examines the legal protection and management of inland waters in Europe.

With inland waters facing a range of significant threats, including from climate change, pollution, and overexploitation, this volume discusses legal solutions for protecting and managing these inland waters. It presents comparative studies from a range of European countries, including the Czech Republic, Germany, France, Lithuania, Poland, and Spain, with each chapter examining the legal status and legal instruments being implemented in each country, allowing for easy cross-comparison. Topics covered include surface water management and protection, water quality and monitoring, management of hazardous substances, and crisis management. Through these unique comparative studies, the volume highlights legal solutions which can be adopted by other nations and used as models for wider implementation across Europe. In doing so, this book offers important recommendations for changing water law to ensure a water secure future in Europe.

This book will be of great interest to students and scholars of water law, water resource management, environmental policy, and environmental conservation.

**Jarosław Dobkowski** is a Professor in the Faculty of Law and Administration at the University of Warmia and Mazury in Olsztyn, Poland.

**Elżbieta Zębek** is a full Professor, the Faculty of Law and Administration at the University of Warmia and Mazury in Olsztyn, Poland.

**Jakub Jan Zięty** is a Professor in the Faculty of Law and Administration at the University of Warmia and Mazury in Olsztyn, Poland.

# Earthscan Studies in Water Resource Management

**Desalination and Water Security**
*Chris Anastasi*

**Flood Risk and Community Resilience**
An Interdisciplinary Approach
*Lindsey Jo McEwen*

**Water Politics and the On-Paper Hydropower Boom**
Power, Corruption, and Sustainability in Emerging Economies
*Özge Can Dogmus*

**Water Justice and Groundwater Subsidies in India**
Equitable and Sustainable Access and Regulation
*Gayathri D. Naik*

**Climate Change and Water Scarcity in the Middle East**
A Transitional Approach
*Marielle Snel, Nikolas Sorensen, and Reed Power*

**Urban Water Ecosystems in Africa and Asia**
Challenges and Opportunities for Conservation and Restoration
*Edited by Shamik Chakraborty, Amit Chatterjee, and Pankaj Kumar*

**Public Banks and Public Water in the Global South**
Financing Options for Sustainable Development
*Edited by Thomas Marois, David A. McDonald and Susan Spronk*

**Legal Protection of European Inland Waters**
Solutions for Sustainable Management
*Edited by Jarosław Dobkowski, Elżbieta Zębek and Jakub Jan Zięty*

For more information about this series, please visit: www.routledge.com/books/series/ECWRM/

# Legal Protection of European Inland Waters

## Solutions for Sustainable Management

Edited by Jarosław Dobkowski,
Elżbieta Zębek, and Jakub Jan Zięty

Routledge
Taylor & Francis Group
LONDON AND NEW YORK

earthscan
from Routledge

First published 2025
by Routledge
4 Park Square, Milton Park, Abingdon, Oxon OX14 4RN

and by Routledge
605 Third Avenue, New York, NY 10158

*Routledge is an imprint of the Taylor & Francis Group, an informa business*



*British Library Cataloguing-in-Publication Data*
A catalogue record for this book is available from the British Library

*Library of Congress Cataloging-in-Publication Data*
Names: Dobkowski, Jarosław, editor. | Zębek, Elżbieta, editor. | Zięty, Jakub Jan, editor.
Title: Legal protection of European inland waters : solutions for sustainable management / edited by Jarosław Dobkowski, Elżbieta Zębek, and Jakub Jan Zięty.
Description: Abingdon, Oxon [UK] ; New York, NY : Routledge, 2025. |
Series: Earthscan studies in water resource management |
Includes index.
Subjects: LCSH: Fresh water--Law and legislation--European Union countries. | Water resources development--Law and legislation--European Union countries. | Water resources--European Union countries--Management. | Water--Pollution--Law and legislation--European Union countries.
Classification: LCC KJE6094 .L44 2025 (print) | LCC KJE6094 (ebook) | DDC 343.409/24--dc23/eng/20240905
LC record available at https://lccn.loc.gov/2024041191
LC ebook record available at https://lccn.loc.gov/2024041192

ISBN: 978-1-032-95866-8 (hbk)
ISBN: 978-1-032-95867-5 (pbk)
ISBN: 978-1-003-58691-3 (ebk)

DOI: 10.4324/9781003586913

Typeset in Sabon
by SPi Technologies India Pvt Ltd (Straive)

# Contents

# Contributors

**Richard Bartes** is an assistant professor at Brno University of Technology, Faculty of Civil Engineering, doctor of legal sciences, Institute of Construction Economics and Management in Brno, Czech Republic. He specialises in the public law (especially in the financial and administrative law) and comparative law with a specialisation in French law. Author of more than 40 reviewed publications in different journals and books. He is a member of Association Henri Capitant des Amis de la Culture Juridique Française.

**Claudia Conradi** is a master of law, lecturer at Saxon University of Applied Administrative Sciences (UAAS) (Germany), deputy head of the Faculty General Administration. Lecturer for constitutional and administrative law, Inspector for environmental concerns (specialised in law and technologies). Author of publications in the field of administrative law.

**Jarosław Dobkowski** is habilitated doctor of legal sciences, professor at the University of Warmia and Mazury in Olsztyn, dean of the Faculty of Law and Administration UWM University of Warmia and Mazury in Olsztyn (Poland), head of the Department of Administrative Law and Safety Sciences UWM in Olsztyn. Author of more than 250 reviewed publications in different journals and books. He is a member of Education Association of Public Administration, administrative law practitioner and member of the Local Government Appeals Court in Olsztyn.

**María José Cazorla González** is a full professor of civil law at the University of Almeria (Spain). She is a member of the UMAU and the CEDR internationally, and of research journals as an external evaluator or as a member of the scientific committee in national and international journals. She is the main researcher in R&D&I and international cooperation, and a full-time researcher in agricultural projects, vulnerable groups, insolvencies, and consumption, as well as the researcher responsible for the Spanish team in the European PSEFS project and researcher in other international projects.

**Michał Mariański** is habilitated doctor of legal sciences, professor at the University of Warmia and Mazury in Olsztyn, Department of Financial Law and Tax Law UWM in Olsztyn, Poland. Graduated at Faculté de droit – Université d'Auvergne Clermonf-Ferrand (France). He specialises in the law of financial market, international private law and comparative law with a specialisation in French law. Author of more than 70 reviewed publications in different journals and books. Member of Association Henri Capitant des Amis de la Culture Juridique Française and an attorney at law being a member of Regional Bar Associations in Olsztyn.

**Kristina Mikalauskaitė-Šostakienė** is a lecturer at Mykolas Romeris University, Faculty of Public Security, Department of Law, Kaunas (Lithuania). She specialises in the public law, legal fundaments and criminal law, especially human rights and democracy, theory of law, law enforcement, business law, human rights, administrative law, environmental law. Author of publication in the field of environmental law (spatial planning, waste, waters).

**Janusz Orłowski** is a doctor of laws, assistant professor in the Department of Financial Law and Tax Law at the Faculty of Law and Administration of the University of Warmia and Mazury in Olsztyn (Poland). He specialises in tax law and public finance law, procedures and forms of operation of the tax administration. He is a legal advisor, appointed civil servant, and has 30 years of work experience as a lawyer. Author and co-author of over 80 academic publications, including those published in scientific journals and conference proceedings in Germany, Czechia, Ukraine, Italy, and the USA, including 1 monograph and co-authorship of 7 collective works.

**Veronika Tomoszková** is an assistant professor at Palacký University Olomouc, Faculty of Law, Department of Administrative and Financial Law (Czech Republic). She specialises in the environmental law (on the national, European Union, international levels) and researches it in the context of other areas of public law (administrative and constitutional law) as well as in comparative perspective. Author of more than 40 reviewed publications in different journals and books.

**Elżbieta Zębek** is a full professor of environmental engineering sciences University of Warmia and Mazury (Poland), The Faculty of Law and Administration, Department of Economic Law and Company Law, head of the Legal and Environmental Laboratory. She specialises in the environmental law (water law, waste management, nature protection), ecology of waters (effects of lake restoration, rivers, climate change on Antarctic), and environmental protection (water management and protection, waste management). Author of more than 150 publications, including articles in different journals, 8 books and 2 scientific editing of monographs.

**Dominik Zidek** is an assistant professor at the Department of Environmental and Land Law at the Masaryk University Faculty of Law (Czech Republic). He professionally specialises mainly in public construction law processes in relation to the protection of environmental interests. From 2014 to 2017 – an assistant judge at the administrative section of the Regional Court in Brno, and since January 2018 – an assistant judge at the Constitutional Court of the Czech Republic. He is also lecturer for the professional public.

**Jakub Jan Zięty** is habilitated doctor of legal sciences, professor of university, head of the Department of Economic Law and Commercial Law at the Faculty of Law and Administration of the University of Warmia and Mazury in Olsztyn (Poland), and associate at the Center for Bioeconomy and Renewable Energy of the University of Warmia and Mazury in Olsztyn. Author of publications in the field of corporate law, real estate law, local government law, legal adviser running his own practice.

# Abbreviations

| | |
|---|---|
| *BayWG* | Bayerisches Wassergesetz (Germany) |
| *BbgWG* | Brandenburgisches Wassergesetz (Germany) |
| *BGBl* | *Journal of Laws* (*Bundesgesetzblatt*, Germany) |
| *BremWG* | Bremisches Wassergesetz (Germany) |
| *BWG* | Wassergesetz (Berlin, Germany) |
| *CAT* | Control and Audit Team (Poland) |
| *CLB* | Central Laboratory of GIEP (Poland) |
| *CMC MCE* | *Crisis Management Center of the Ministry of Climate and Environment (Poland)* |
| *CMI* | Central Mining Institute (Poland) |
| *CZK* | Czech currency Koruna |
| *CzWL* | Czech Water Law (Act No. 254/2001 Coll.) |
| *DEM* | Department of Environmental Monitoring (Poland) |
| *DVBl* | German Administrative Gazette (Deutsches Verwaltungsblatt, Germany) |
| *EAP* | 8th Environment Action Programme until 2030 |
| *ECHA* | European Chemicals Agency |
| *EC* | European Commission |
| *EEA* | European Environmental Agency |
| *EIA* | Environmental Impact Assessment |
| *E. n* | Edge number (Randnummer, Germany) |
| *EPTB* | Territorial public establishments for the river basin (Fr. établissements publics territoriaux de basin, France) |
| *EQSD* | Directive 2008/105/EC on Environmental Quality Standards |
| *EU* | European Union |
| *FRS C* | Fire Rescue Service of the Czech Republic |
| *GDEP* | General Director of Environmental Protection (Poland) |
| *GG* | The Basic Law of the Federal Republic of Germany (*Grundgesetz*) |
| *GIEP* | General Inspector for Environmental Protection (Poland) |
| *HWaG* | Hamburgisches Wassergesetz (Germany) |
| *HWG* | Hessisches Wassergesetz (Germany) |
| *INSPIRE* | Directive 2007/2/EC |

| | |
|---|---|
| *IWM* | Integrated Management Frameworks |
| *LWaG* | Wassergesetz des Landes Mecklenburg-Vorpommern (Germany) |
| *LWG* | Landeswassergesetz (Germany) |
| *MAPTAM* | Act No. 2014-58 of 27 January 2014 on modernisation of territorial public actions and affirmation of a metropolis (France) |
| *NVwZ* | New Journal for Administrative (Neue Zeitschrift für Law Verwaltungsrecht, Germany) |
| *NWG* | Niedersächsisches Wassergesetz (Germany) |
| *PCMC* | Provincial Crisis Management Center (Poland) |
| *PIEP* | Provincial Inspector of Environmental Protection (Poland) |
| *RBD* | River Basin Districts |
| *REACH* | Regulation 1907/2006 |
| *REFIT* | Regulatory Fitness and Performance |
| *SächsWG* | Sächsisches Wassergesetz (Germany) |
| *SDGs* | Sustainable Development Goals |
| *SWG* | Saarländisches Wassergesetz (Germany) |
| *TFEU* | Treaty on the Functioning of the European Union |
| *ThürWG* | Thüringer Wassergesetz (Germany) |
| *WFD* | Water Framework Directive (Directive 2000/60/EC) |
| *wfe* | with further evidence (mit weiteren Nachweisen, Germany) |
| *WG LSA* | Wassergesetz für das Land Sachsen-Anhalt (Germany) |
| *WHO* | World Health Organization |
| *WHG* | German law in Gesetz zur Ordnung des Wasserhaushalts (*Wasserhaushaltsgesetz*) |
| *WG* | Wassergesetz für Baden-Württemberg (Germany) |
| *WL* | Water Law of 20 July 2017 (Poland) |
| *UN* | United Nations |

# Introduction

*Elżbieta Zębek*

Water deficit and its deteriorating quality have recently become a topical issue. This is caused by the progressing climate change and excessive anthropopressure, which leads to the formation of semi-desertified areas with water deficit, which is particularly true for the Mediterranean region of the European Union (EU), as well as to excessive water pollution, which restricts its use for household and economic purposes. According to a report prepared by the European Environmental Agency (EEA), countries of Southern Europe are still struggling with a huge water deficit, which also entails a deficit in the north of the continent. Moreover, water resources are exploited excessively, which has a strong impact on the quality and quantity of the remaining water and on ecosystems. The water resources are disconcertingly small (under 3,000 $m^3$/person), with those in Poland, Malta, Cyprus, and Czechia being below the water safety level. According to the United Nations (UN), 1,700 $m^3$ per person is the limit below which a country is regarded as threatened with a water deficit (EEA Report No. 2/2009). Hence, there is a need for special legal protection of water resources. Proper water resources management plays a key role in ensuring sustainable water use and in preserving the world's natural capital (UN Resolution No. 2187/2021).

In the protection of surface waters in the world, of great importance was the Convention on the Law of the Non-navigational Uses of International Watercourses adopted by the General Assembly of the UN on 21 May 1997, which was considered a milestone in the evolution of international water law (Eckstein, 2021). This convention applies to the use of international watercourses and their waters for purposes other than navigation and related to measures for the protection, preservation, and management of the use of these watercourses and their waters. The issue of inland water protection is also included in the Convention on the Protection and Use of Transboundary Watercourses and International Lakes, signed in Helsinki on 17 March 1992, to which European countries are parties. The act was enacted due to the need for urgent protection of transboundary watercourses and international lakes, the effective implementation of which can only be ensured by close cooperation. For this purpose, measures should be taken to prevent and reduce the discharge of hazardous substances into the aquatic environment and to reduce

DOI: 10.4324/9781003586913-1

the eutrophication of surface waters. Under Article 2, states parties are obliged to take appropriate measures to prevent, control, and reduce any transboundary impact by preventing, controlling, and reducing water pollution, as well as ensuring the use of such waters in a reasonable and equitable manner consistent with ecologically sound and rational water resources management, conservation and environmental protection, and the restoration of aquatic ecosystems. The following principles should be followed in implementing these projects: (a) the precautionary principle aimed at avoiding potential transboundary pollution caused by the discharge of hazardous substances even when a cause-and-effect relationship between these substances and potential transboundary impacts has not been definitively established; (b) the polluter pays principle, according to which expenses related to pollution prevention and reduction are borne by the polluter; and (c) management of water resources in such a way that meeting the needs of the present generation does not reduce the ability of future generations to meet their needs. The convention's overall goal in Article 3 is to prevent, control, and reduce water pollution. Appropriate legal, administrative, economic, financial, and technical measures should be introduced in this regard, which include (a) the use of low- and zero-waste technologies; (b) permits for wastewater discharges by competent national authorities and monitoring and control of such discharges; (c) the setting of limit standards for the discharge of hazardous substances; (d) the use of at least biological treatment of urban wastewater; (e) the development and application of appropriate measures and ecological best practices, leading to the reduction of nutrients and hazardous substances from diffuse sources; (f) the carrying out of aquatic environmental impact assessments; (g) the development of plans to counter emergency hazards; and (h) the reduction of the risk of emergency pollution. The provisions of this convention are included in the EU legal regime for the protection of surface water.

Resolution 2020/2613 emphasises that surface waters, especially rivers, are the most endangered areas, even though they are considered the largest providers of ecosystem services because they enable the development of economic activities. Member states should therefore adopt measures to reduce their pollution. In addition, implementing measures that are necessary to achieve good status of water bodies should be identified, and action programmes should be developed based on the best available evidence. Public access to information regarding such programmes and surface water impact assessments should also be improved. EU action has led to a reduction in chemical pollution of EU waters. However, there is an urgent need to update the list of priority substances. The European Commission (EC) should take all necessary measures to achieve good chemical status and take decisive EU-wide action if member states fail to meet environmental quality standards for priority substances that fall within the scope of EU legislation. Monitoring systems for water quality and environmental pollution should also be simplified and improved, collecting data on major sources of emissions of hazardous substances, among other things.

The system of surface water legal protection in the EU is based on the Water Framework Directive No 2000/60/EC (WFD). The purpose of this directive is to establish the framework for all the types of waters covered by the regulation, which (a) prevent further deterioration and protect and improve the status of aquatic ecosystems and – with respect to their water needs – land ecosystems and wetlands directly dependent on aquatic ecosystems; (b) promote sustainable water use based on long-term protection of available water resources; (c) strive for increased protection and improvement of the aquatic environment by specific measures designed to achieve gradual reduction of discharges, emissions and loss of priority substances and discontinuation or gradual elimination of discharges, emissions and loss of priority hazardous substances; (d) ensure gradual reduction of groundwater pollution and prevent its further pollution, and (e) contribute to reduction of the impact of floods and droughts (Art. 1). Therefore, according to the WFD environmental objectives, it is necessary to step up protection and improve the aquatic environment quality, as well as to gradually reduce the emission of priority hazardous substances (Ciechanowicz-McLean, 2013). Therefore, these objectives stem from supplementing the policy with the sustainable development principle, and they involve satisfying the water requirements of people, agriculture, and industry, promoting sustainable water use, protecting waters and ecosystems in good ecological condition as well as improving the quality of water and ecosystems degraded by human activity. Under Article 4, member states were required to take the necessary measures for surface waters to prevent deterioration of the status of all bodies of surface water, both natural and artificial and heavily modified. They should also take measures to progressively reduce pollution substances and to cease or gradually reduce emissions, discharges, and losses of priority hazardous substances. The WFD thus reflects new directions in water management, taking into account the basic principles of environmental principles of environmental law, and it has the following main objectives: an integrated EU policy for the long-term sustainable use of water while respecting the principle of subsidiarity, achieving 'good status' for all waters by 2015, water management based on a 'combined approach' for river based on river basins with a 'combined approach' of emission limits and quality standards, price regulation for water use to ensure that the polluter pays principle, and more effective water protection legislation (Stoa, 2014).

The WFD is widely accepted and recognised as the most important piece of European water legislation (Voulvoulis et al., 2017). However, according to the EC reports, some WFD provisions are too restrictive and difficult for member states to implement. The most important issues include (a) a lack of sufficient tools for environmental and economic assessment; (b) a gap in specific drainage basins or specific water categories, also those used for consumption, especially in countries that accessed the EU in 2004 and 2007; (c) insufficient budget in the member states to attain the WFD objectives by 2015, especially for some important 'basic measures', e.g., wastewater treatment or resolving the issue of nitrate emission from agriculture; (d) monitoring programmes, which focused on 'examining specific structural parameters, assuming that good quality of such elements corresponds to the good functioning of ecosystems'; therefore, the programmes focus on symptoms rather than on causes for water degradation (Błaszczak, 2005; Maciejewski & Walczykiewicz, 2006; Pawlaczyk, 2012; Bouleau & Pont, 2015; Giakoumis & Voulvoulis, 2018; Zębek, 2022). In addition, problems in implementing these provisions of the WFD in some European countries may be caused by the fragmentation of water management bodies because responsibilities are highly fragmented between them. Other agencies are responsible for management, monitoring, and water quality. There is no complete database or understanding of the links between the existing multi-level regulatory institutions in these countries. Water quality standards in these countries are regulated for a specific type of water use in water law (Vystavna et al., 2018). The EC report on the execution of the water framework directive (2000/60/WE) and the flood directive (2007/60/WE) notes that given the climate changes, managing such an important resource as water requires an integrated approach, which takes into account the environmental, social, economic, and health-related dimension. This report shows the progress of the WFD implementation, especially with respect to achieving a good condition of surface waters by 2015, with the directive allowing for extending the period for up to a maximum of two further cycles (i.e., the present one of 2015–2021 and a next one of 2021–2027). All member states approved their water management plans for the drainage basins and flood risk management plans, except for the Canary Islands (Spain). This information was included in the digital database – the European Water Information System, maintained by the EEA, which was the basis for the assessment. According to the report, only 38% of the surface waters are in good chemical status, and a good ecological status/potential was noted for barely 40%. Compared with the 2009–2015 cycle, only a small portion of surface waters recorded improved quality, which could be caused by a late identification of impact, a longer period required for developing effective measures in terms of policy, slow introduction of measures, the time of nature's response before the measures trigger the effects, and the strict quality standards and improvement of monitoring and reporting, which shows that the condition of water bodies which was previously regarded as 'unknown' was, in fact 'not satisfactory'. In addition, the EC has highlighted the problematic issue of impacts from industrial sectors in terms of pollutants resulting in failure to meet the requirements for chemical or ecological status. The EC concludes that further progress of work in this direction is needed, assessing as insufficient the measures introduced to aimed at limiting or stopping the discharge into waters of certain pollutants (Woźnicka & Kuczyńska, 2019). The EEA report also states that European waters are strongly affected by both dispersed (e.g., agriculture, transport infrastructure) and spot sources (e.g., industry or energy generation) of pollutant emission, excessive water intake, and hydromorphological changes resulting from a range of human activities. Monitoring water quality elements is incomplete, and it is based excessively on grouping a range of various water bodies and on expert opinions rather than on a more detailed assessment of each respective water body in accordance with the parameters contained in the water framework directive. Further efforts are needed to make the appropriate monitoring networks reach a sufficient spatial range and assessment credibility. However, the reporting indicates that compliance with the water framework directive objective is gradually increasing. Although appropriate policy measures have been applied in many member states and a range of financial investments have been made, water quality improvement will be delayed. Although less than half of the surface waters are in a good status, the tendency regarding a range of basic quality elements and those concerning substances is more positive. Nevertheless, the European Parliament's report 2020 shows that the WFD is fit for purpose. Implementation must be improved and expedited. While an Integrated Management Framework (IWM) for many thousands of water bodies in the thousands of water bodies in the EU helps to slow the rate of water and reduce chemical pollution, less than half of the EU's water bodies are in good status, despite the 2015 deadline. Furthermore, only one of the four freshwater indicators has improved over the last 10–15 years. This is due to problems with the implementation of water legislation, mainly due to insufficient funding and inadequate environmental objectives in environmental sector policies (European Parliament, 2020).

Due to the previously indicated problems with the implementation of the Water Framework Directive (WFD), resulting in insufficient improvements in surface water quality in EU countries, there is a need for studies indicating how its requirements are implemented by selected countries. The WFD provisions were implemented in the national legislation of EU member states, including Czechia, France, Spain, Lithuania, Germany, and Poland. The system of legal water protection in these countries is, therefore, based on WFD provisions, the main objectives of which are identical to those of all EU countries. This study analyses the systems of legal protection of surface waters in selected EU countries, i.e., Czechia, France, Spain, Lithuania, Germany, and Poland. The aim of the analysis is to present the legal status and instruments of surface water management and protection with an assessment of their effectiveness in these countries. The countries selected for analysis are members of the EU, but they joined this international organisation in different years. France and Germany are founding members, Spain joined in 1985, and Poland, the Czech Republic, and Lithuania joined in 2004. The latter countries therefore acceded after the WFD had been in force for four years, which could undoubtedly have influenced the process of implementing its provisions. These countries are characterised by peculiar legal solutions in this regard depending on different political, geographic, and environmental conditions. The level of environmental policy in these countries depends on many factors, including the size of the country, natural conditions, and the financial resources spent on environmental protection. The largest of the analysed countries in terms of area is France, also in the EU, followed by Spain, Germany, and Poland, while the smallest are the Czech Republic and Lithuania (EU, 2024). These countries have varying levels of surface water resources, ranging from 1.5 in the Czech Republic, 1.6 in Poland, 3.3 in Germany and Spain, 3.1 in France to 7.9 thousands of $m^3$/inhabitant in Lithuania (GUS, 2024). These countries also differ in their status with regard to failure to achieve good surface water chemical status in terms of priority substances, i.e., the Czech Republic –30%, Germany –8%, France –20%, Poland –30%, and Spain –8% (EEA, 2018). Germany is one country where the share of the population connected to at least secondary treatment will rise to 95% by 2022; the Czech Republic also ranks high; Spain and France are above 80%, and Lithuania and Poland are at 76%. Unless the receiving water is in a sensitive area, the minimum secondary treatment plant will normally provide an acceptable level of environmental protection (Eurostat, 2022). The measure of financial expenditure on environmental protection can be environmental protection transfers by environmental protection activity and institutional sector. Here, the highest expenditure in 2021 will be in Germany and the lowest in Lithuania and the Czech Republic, while the rest of the countries analysed will be at an average level (Eurostat, 2021).

In each of the analysed countries in the water law, on the one hand, similar legal instruments have been applied (the system of water permits, planning, monitoring), and on the other hand, due to local conditions and problems, appropriate systems of monitoring and analysis of risks associated not only with flood conditions but also severe water pollution have been implemented (e.g., Czech Republic, Poland, Germany). Indeed, the experience of these countries, especially the effective instruments and legal solutions applied, can help develop a universal system of legal protection of surface water in the EU. Identification of legal gaps in this system can guide legislative and organisational changes aimed at improving surface water quality not only in the countries analysed, but it can also be implemented in other member states. Therefore, these comparative studies allowed the authors to develop recommendations for changes in water law that could be adopted in many countries. The analysis conducted on the necessity of changes in water law is of particular importance for the preservation of water security in the EU. The added value of the book lies in the recommendations for water law in order to achieve more effective protection of inland waters in the EU. This study employed the legal-dogmatic and the legal-comparative method. The chosen research method makes it possible to compare the water law systems in the analysed countries. For this, the same issues were analysed for each country in terms of general legal principles of water management and protection, the structure and role of water authorities, surface water quality monitoring and assessment, and protection of water quality via regulation of wastewater, hazardous substances, and accidents, as well as administrative and penal responsibility of entities using surface waters. Thus, thanks to the comparative analysis of the same issues, the monograph constitutes a coherent whole and allows the formulation of relevant conclusions that constitute recommendations for changes in the legal system of management and protection of EU surface waters, which is the primary goal of this study.

## References


Błaszczak, T. (2005). Ramowa Dyrektywa Wodna: Strategia wdrażania. In *Integrated Coastal Zone Management at the Szczecin Lagoon: Exchange of experiences in the region*, edited by Sekścińska & Löser, 6. Coastline Reports.

Bouleau, G., and Pont, D. (2015). Did You Say Reference Conditions? Ecological and Socioeconomic Perspectives on the European Water Framework Directive. *Environmental Science and Policy* 47: 32–41. https://doi.org/10.1016/j.envsci.2014.10.012

Ciechanowicz-McLean, J. (2013). Ramowa Dyrektywa Wodna a ochrona środowiska morskiego. *Prawo Morskie* 29: 107–115.

Eckstein, G. (2021). The status of the UN Watercourses Convention: does it still hold water?, 1 ed. Global Water Resources, Routledge.

EEA. (2009). Report No. 2/2009, Water resources across Europe — confronting water scarcity and drought EEA. Copenhagen, https://www.eea.europa.eu/publications/water-resources-across-europe

EEA. (2018). Report No 7/2018, European waters assessment of status and pressures 2018. Luxembourg.

EU. (2024, August 20). Fakty i liczby dotyczące życia w Unii Europejskiej, 2024. https://european-union.europa.eu/principles-countries-history/key-facts-and-figures/life-eu_pl

European Parliament. (2020). Implementation of EU water legislation, Procedure: 2020/2613 (RSP), Interpellations of 7 December 2020. https://www.europarl.europa.eu/doceo/document/O-9–2020–000078_EN.html

Eurostat. (2021, August 21). Eurostat, Environmental protection transfers by environmental protection activity and institutional sector, 2021. https://ec.europa.eu/eurostat/databrowser/view/env_ac_eptrf1/default/bar?lang=en

Eurostat. (2022, August 20). The share of the population connected to at least secondary wastewater treatment plant, 2022. https://ec.europa.eu/eurostat/statistics-explained/index.php?title=Water_statistics#Water_as_a_resource

Giakoumis, T., and Voulvoulis, N. (2018). Progress with monitoring and assessment in the WFD implementation in five European River Basins: significant differences but similar problems. *European Journal of Environmental Sciences* 8(1):44–50. https://doi.org/10.14712/23361964.2018.7

GUS. (2024). *Ochrona środowiska 2023*. Warsaw.

Maciejewski, M., and T. Walczykiewicz (2006). Dotychczasowe doświadczenia związane z wdrażaniem Ramowej Dyrektywy Wodnej. *Infrastruktura i Ekologia Terenów Wiejskich* 4(1): 63–76.

Pawlaczyk, P. (2012). Perspektywy osiągnięcia celów Ramowej Dyrektywy Wodnej w Polsce. *Przegląd Przyrodniczy* 23(3): 52–68.

Stoa, R. (2014). Subsidiarity in Principle: Decentralization of Water Resources Management. *Utrecht Law Review* 2: 35. https://doi.org/10.18352/ulr.267

Voulvoulis, N., K.D. Arpon, and T. Giakoumi (2017). The EU Water Framework Directive: From great expectations to problems with implementation. *Science of the Total Environment* 575:358–366. https://doi.org/10.1016/j.scitotenv.2016.09.228

Vystavna, Y., M. Cherkashyna, and M.R. van der Valk (2018). Water laws of Georgia, Moldova and Ukraine: current problems and integration with EU legislation. *Water International* 43(3), 424–435. https://doi.org/10.1080/02508060.2018.1447897

Woźnicka, M., and A. Kuczyńska (2019). Revision of EU directives on water resources management – plans of the European Commission regarding the amendment of EU water policy. *Biuletyn Państwowego Instytutu Geologicznego* 475: 237–244, https://doi.org/10.7306/bpig.29

Zębek, E. (2022). Legal solutions of lake monitoring systems in Poland in compliance with the Water Framework Directive. *Review of European and Comparative Law* 48(2): 173–201. https://doi.org/10.31743/recl.13419

# 1 The protection of inland waters in the EU

*Elżbieta Zębek*

## EU environmental policy and water protection

The protection of inland waters in the European Union (EU) has a special place in the 8th Environment Action Programme to 2030 (EAP; EU Decision 2022/591), which entered into force on 8 May 2022. This policy is based on Articles 11 and 191–193 of the Treaty on the Functioning of the European Union. Article 191 of Title XX, entitled 'Environment', states that the Union policy on the environment shall contribute to (a) preserving, protecting, and improving the quality of the environment, (b) protecting human health, (c) careful and rational utilisation of natural resources, and (d) promoting measures at international level to deal with regional or worldwide environmental problems, and in particular combating climate change. Furthermore, the principles of environmental law, i.e., sustainable development, prevention, precaution, rectification of damage at source, and the polluter-pays principle, underpin the implementation of these objectives. This programme provides a framework for the overall development of EU environment policy, including water. The EAP aims to ensure the effective implementation of EU climate and environmental legislation, which underpins the implementation of the United Nations (UN) Agenda 2030 and its Sustainable Development Goals (SDGs). The UN Agenda 2030 (UN Resolution No. A/RES/70/1) is a road map for people, planet, and prosperity, with 17 SDGs aimed at achieving human rights, including the right to a clean environment (water, air, etc.). These goals are integrated and indivisible, balancing the three dimensions of sustainable development: economic, social, and environmental. When analysing the SDGs, the most important goal related to water management and protection is Goal 6: ensure availability and sustainable management of water and sanitation for all (Pedersen et al., 2023). According to the 8th EAP, action to achieve the EU's environmental and climate change goals must be coherent with the implementation of the European Pillar of Social Rights. This programme identifies priority targets for 2030 and the preconditions for achieving them. This document sets out a framework of six priority objectives, of which the following are of key importance for water protection: (4) achieve zero emissions of pollutants, including air, water, and soil pollution,

DOI: 10.4324/9781003586913-2

and protect the health and well-being of Europeans, and (5) protect, conserve, and restore biodiversity and enhance natural capital, in particular air, water, soil and forests, freshwater, wetlands, and marine ecosystems (see Pindaru et al., 2023; Zębek, 2023b). The priority goals must be achieved by 2030 at the latest. This programme should become a management tool that shapes EU environmental policy. It has introduced a new approach based on the principle of 'sustainability first' (European Parliament, 2021).

The 'Pathway to a Healthy Planet for All EU Action Plan: Towards Zero Pollution for Air, Water and Soil' (COM (2021)400) is a document from the EC that outlines a new strategy to combat pollution. Zero emissions is a cross-cutting goal that contributes to the implementation of the UN 2030 Agenda for Sustainable Development and complements the goal of climate neutrality by 2050, in synergy with the goals of a clean and circular economy and the restoration of biodiversity. The plan is based on the precautionary principle and the principles of preventive action, with the goal of eliminating pollution by first addressing it at its source using the 'polluter-pays' approach. This programme changes the approach, particularly in the area of water protection, where the first step is to identify and eliminate the source of pollution. This document introduced a new hierarchy of actions for the elimination of pollution based on the precautionary principle and the principles of preventive action, 'repairing damage at source' and the 'polluter-pays' principle. This means that pollution should first be prevented at the source.

Where pollution cannot be prevented in its entirety, it should be minimised, as stated in one of the most important principles of environmental law – the precautionary principle. In addition, once pollution has occurred, it should be cleaned up immediately, and the resulting damage should be compensated for. Therefore, the hierarchy of actions to eliminate pollution will consist of reversing the pyramid of actions and prioritising approaches to pollution control, i.e., eliminating existing pollution as far as possible and taking measures to restore 'good status'. The main objective of this road map is to provide guidance on how to mainstream pollution prevention measures into all relevant EU policies, maximise synergies in an effective and proportionate way, accelerate their implementation, and identify possible gaps or trade-offs. The road map sets out key targets for 2030 to accelerate the reduction of pollution in order to guide the EU towards the vision of a healthy planet for all by 2050. The plan prioritises anti-pollution measures, such as eliminating existing contamination of water and soil and working to restore them to a 'good condition'. To this regard, the programme will review and update water legislation, in particular to better adapt it to reduce chemical pollution and microplastics, and will also help preserve the quality of drinking water (EC, 2021; Zębek, 2023a). The aforementioned objectives are the subject of a large body of EU legislation in the field of water management and protection.

The implementation of the European Citizens' Initiative Right2Water (Resolution 2017/C 316/0917) is important to prevent water shortages in the EU and to improve access to and quality of drinking water. The main message of the campaign was that EU legislation should oblige national governments to ensure that all citizens have access to sufficient clean drinking water and sanitation. The commission conducted an EU-wide public consultation and a Regulatory Fitness and Performance Evaluation (REFIT) of Directive 98/83/EC. This directive established a legal framework for the protection of human health against the adverse effects of any contamination of water intended for human consumption, ensuring that it is wholesome and clean. The European Commission intended to act in several areas related to this initiative (increasing transparency, stimulating innovation, etc.) and presented a legislative proposal for the revision of the Drinking Water Directive, including an obligation for Member States to ensure access to water for the most vulnerable. The European Commission has a proposal for a revision of the rules on the status of and access to and access to drinking water. The new rules aim to ensure high-quality tap water across the EU. This initiative led to the adoption of Directive (EU) 2020/2184 on the quality of water intended for human consumption. This directive will have the same purpose as Directive 98/83/EC and should increase access to drinking water for all in the EU. In line with this objective, this directive aims to protect human health against the harmful effects of any contamination of water intended for human consumption by ensuring that it is wholesome and clean and by improving access to water intended for human consumption (Article 1). It is therefore necessary to establish at Union level minimum requirements to be met by water for this purpose. Member States must take the necessary measures to ensure that water supplied for human consumption is free from all microorganisms and parasites and from all substances which, in their numbers or in their concentrations, are potentially harmful to human health and that it meets these minimum requirements (Article 4). This directive addresses the need to identify hazards that may affect drinking water supplies and to monitor pollutants that have been identified as significant, i.e., nitrates, pesticides, or pharmaceuticals and arsenic. Member States should pay particular attention to the content of microplastics in the risk assessment of water intended for human consumption from surface waters (see Zębek, 2023b).

Communications SWD (2019)439 and SWD (2020)259 (EC, 2019a; EC, 2020a) identify that the implementation of freshwater legislation is still insufficient due to factors such as lack of investment, limited integration of freshwater protection objectives into other policy areas, slow implementation of measures, and the need to better address chemical contamination. Actions will therefore mainly focus on strengthening implementation. Member States should therefore promote sustainable and efficient water use by taking measures to reduce water pollution so that consumers and polluters bear the appropriate costs while encouraging sustainable investment. Better monitoring and control of surface water pollution from priority substances should also be supported. In addition, a review of the Urban Waste Water Treatment Directive is foreseen, in synergy with the evaluation of the Sewage Sludge Directive, which will help to raise the level of ambition for nutrient removal from wastewater and the treatment of treated sludge and water for re-use, thus supporting less polluting circular agriculture. The review will also address emerging pollutants, such as microplastics and micropollutants, including pharmaceuticals.

In the EU, the issue of microplastic water pollution has recently been of great concern due to the increase in data on the levels of microplastics in surface waters and their negative impact on the environment and humans (Horton et al., 2017; Karbalaei et al., 2018). Microplastics can enter freshwaters from surface runoff, domestic and industrial wastewater, degraded plastic waste, and atmospheric deposition (Carpenter & Smith, 1972; Dusaucy et al., 2021). The issue of protecting waters from microplastic pollution was addressed in Communication COM (2021)400. One of the targets for zero pollution by 2030 – in line with EU law and the guidelines included in the Green Deal and in synergy with other EU initiatives – is to reduce by 50% the amount of plastic waste in the sea and by 30% the amount of microplastics released into the environment. It should be emphasised that legislative work on the reduction of microplastics in the environment is still ongoing, and the form and effectiveness of such work will be largely dependent on the actions of the European Commission together with the Council and the European Parliament in this area (Kentin, 2018). One of the initiatives aimed at solving the problem of pollution of the environment, including surface waters, with toxic substances, including microplastics, was the adoption by the European Commission in 2018 of a strategy aimed, among other things, at reducing sources of microplastic pollution. As part of these activities, in 2019, the European Chemicals Agency (ECHA) proposed broad restrictions on the use of microplastics in products placed on the EU market, with the aim of limiting the release of microplastics into the environment. These activities led to the development of a draft regulation amending Annex XVII of Regulation 1907/2006 (REACH) in relation to synthetic polymer microparticles. This document included synthetic polymer microparticles, thus extending the qualification of microplastics to chemical substances, mixtures, and their compounds (EC, 2023; Zębek, 2023c).

## Legal principles for the management and protection of inland waters in the EU

In the EU, the main piece of legislation governing water is the Water Framework Directive No. 2000/60/EC (WFD), which establishes a framework for community action in the field of water policy. The main message of this law is that 'water is not a commercial product like any other, but rather a heritage that must be protected, defended and treated as such'. The aim is to provide a framework for the protection of inland surface waters, including those analysed in this study. More specifically, it allows for the protection of these waters, the restoration of ecosystems in and around these waters, the reduction of water pollution, and the sustainable use of water by individuals and businesses. The directive sets rules to halt water degradation and achieve 'good status' of EU rivers and lakes by 2015 (see Ciechanowicz-McLean, 2013; Voulvoulis et al., 2017).

The WFD recommends that inland waters be managed according to the River Basin Management Plans. The European Commission has defined a common implementation strategy, in particular guidelines for monitoring. This strategy aims to promote consistency in the implementation of the monitoring system by providing additional guidance on the design and implementation of chemical and biotic standards, sampling, processing, and presentation of data and their use for compliance assessment. These waters should therefore be characterised in terms of their bioindicators, chemical elements, and hydromorphological parameters. These indices should be compared with those of waters not affected by human activity and then classified into different ecological status categories, with the aim of achieving 'good status' for all waters (Article 4). In this respect, the provisions of the WFD have significantly extended the environmental objectives, including in particular requirements for the control of wastewater discharges, especially industrial wastewater, and a comprehensive monitoring and reporting framework (EC, 2010, 2014; Farmer, 2012; Giakoumis & Voulvoulis, 2018; Zębek, 2022). In addition, the directive recognises that some water bodies may need more time to achieve good status. For this reason, Member States may use a derogation to the natural conditions of the water body and extend the deadline until 2027. The deadline for achieving good status may also be extended if it is technically impossible or disproportionately expensive. Where exemptions apply, the WFD requires Member States to justify and explain in their River Basin Management Plans. An integrated approach to water protection is essential to achieve 'good status' of water bodies. In international law, there is a concept of ecological integrity of water bodies, which is equivalent to the formula for good ecological status of water bodies. Ecological integrity means the natural conditions of water bodies and other resources sufficient to maintain the biological, chemical, and physical integrity of the aquatic environment (Ciechanowicz-McLean, 2013). To achieve this, the first step is to determine the quality status of these water reservoirs through monitoring. Monitoring is an essential part of the implementation of EU water legislation, in particular the WFD (see Vlachopoulou et al., 2014; Bouleau & Pont, 2015; Boeuf et al., 2016; Zębek & Napiórkowska-Krzebietke, 2019).

Three types of monitoring are specified in Annex 5 1.3 of the WFD:

1 Surveillance monitoring is carried out for the following purposes: to provide an overall assessment of the status of surface waters in a given river basin, to complement and validate the assessment of impacts on water ecosystems, to create an effective project of monitoring programmes, and to estimate the programmes, as well as the estimation of long-term changes in natural conditions and their conditions and their changes caused by anthropogenic pressures.

2 Operational monitoring is established mainly to determine the status of water bodies identified as bodies recognised as waters at risk of failing to meet their environmental objectives and to assess changes in the status of such bodies resulting from the measured programmes.
3 Investigative monitoring, referred to as 'supplemental', is conducted in the following cases: inability to identify the cause of the exceedance in water parameters and bioindicators, surveillance monitoring that shows an inability to achieve environmental purposes, operational monitoring that has not revealed the causes of this state, estimation of the extent and impact of accidental pollution, impact of accidental pollution, information about the determined programme of measures to achieve environmental objectives, and special measures necessary to eliminate the effects of accidental pollution.

This monitoring shall be carried out in accordance with the technical standards and requirements for the exchange and collection of information set out in the INSPIRE Directive (2007/2/EC). In addition, Article 8 of the WFD includes the sufficient quality and comparability of the analytical results laboratories for monitoring the ecological and chemical status of waters in accordance with Directive 2009/90/EC. In order to assess the ecological status, indicators of ecological quality are used for a given surface water category. The monitoring elements must be indicative of the pressures operating in the river basin (EC, 2009; See Zębek, 2017; Zębek & Napiórkowska-Krzebietke, 2019).

In order to accomplish the objectives of the WFD, each Member State is obliged to develop a programme of measures for all the drainage basins to accomplish the environmental objectives. Such measures include a ban on introducing pollutants to water for discharges from spot sources which can cause pollution or prior permit or registration based on generally binding rules, which establish emission control for specific pollutants, which are periodically reviewed and updated as needed, as well as other measures for preventing considerable loss of pollutants from technical installations (Article 11). Each programme of measures includes 'basic' and 'additional' measures. The 'basic measures' are the minimum requirements to be met and consist of measures required for the implementation of community legislation on water protection, including measures required under the legislation set forth in Part A of Annex VI. Additional measures should promote the efficient and sustainable use of water to avoid compromising environmental goals. With regard to water pollution, such measures are aimed at preventing or controlling the introduction of pollution from diffuse sources that may cause pollution. Controls can take the form of requiring prior regulation, such as prohibiting pollutants from entering the water without first obtaining a permit. These controls are reviewed periodically and updated as necessary.

**Legal regulations on priority substances and industrial wastewater treatment**

The WFD introduces the requirement for creating a list of priority substances which pose a significant hazard to the aquatic environment at the EU level. The directive also indicates a system for monitoring the condition of surface water as an instrument of water legal protection in the EU. In this regard, the Member States are obliged to develop programmes for monitoring water status in order to establish a coherent and comprehensive overview of the water status in each drainage basin for surface waters, which covers the volume and level or flow intensity within the range suitable for the ecological and chemical condition, ecological potential as well as for the ecological and chemical condition and the ecological potential (Article 8). For such substances, specific measures should be taken against water pollution by these substances, taking into account all significant sources and identifying economically effective and proportionate levels and combinations of controls. The identification of priority hazardous substances should take into account the precautionary principle, which is based, in particular, on determining the potential adverse effects of the product's use and on a scientific risk assessment. Moreover, in order to provide additional protection and improvement of the water status, Member States can adopt further measures mentioned in Assumption 6 of Part B of the directive, which include legal instruments, administrative instruments, and emission controls. This study analyses the solutions and legal instruments implemented in member countries under the provisions of the WFD, which mainly include planning, the water permit system, the duties and powers of the water authority, surveillance and monitoring, and emergency management of uncontrolled wastewater discharges.

To strengthen the protection of surface waters from the effects of toxic and persistent water pollution, the European Commission has identified the need to revise the list of pollutants and quality standards set out in the WFD and the Directive 2008/105/EC on Environmental Quality Standards (EQSD). The proposal introduces a number of amendments and changes aimed at addressing the shortcomings of the Fitness Check of European water legislation and adapting the legal framework to scientific and technical progress. The commission's proposal significantly extends the list of priority substances in the WFD to 68 substances and groups of substances. The revision of the main legislative framework creates opportunities to improve the management of water pollution in line with the 'one substance, one assessment' principle (EC, 2022a; Backhaus, 2023). The necessity of updating these lists was confirmed by the 2019 Fitness Check, which also concluded that other enhancements to the legislation would increase its effectiveness, efficiency, and coherence. Within the overarching objective of EU water policy, the general objectives of the initiative are (1) to enhance the protection of EU citizens and natural ecosystems in accordance with the EU Biodiversity Strategy 2030

(EC, 2020b; EEA 2020; Hermoso et al., 2022) and the EU Pollution Action Plan included in the European Green Deal (EC, 2019b), and (2) to increase the efficiency and reduce the administrative workload of the legislation, enabling the EU to more rapidly respond to emerging threats. The setting of environmental quality standards for chemicals in surface waters and their control is intended to be complementary to the provisions on sources and modes of transport, with the aim of introducing more restrictive production and emission standards. The specific objectives of the initiative are (a) to update the lists of pollutants impacting on surface waters, (b) to improve the monitoring of chemical compounds, (c) to standardise the elimination of pollutants in surface waters in the EU, (d) to ensure that the legislative framework can be updated more quickly in the light of scientific knowledge in order to deal more effectively with emerging pollutants, and (e) to improve access to data, transparency, and re-use in order to increase adherence, reduce administrative workload and improve consistency with the wider EU chemicals regulatory framework. The final objective of the initiative is to set new standards for a number of chemicals of 'potential concern' to control chemical pollution of water, to facilitate implementation on the basis of a simplified and more cohesive regulatory framework, and to provide dynamic and up-to-date information on the status of water.

In the EU, the WFD does not refer directly to microplastics in relation to toxic substances, including man-made polymers in the protection of inland waters. However, EU Member States are required to monitor potential pressures from anthropogenic factors. These are defined in Article 2(29) of the directive as substances that are toxic, persistent and liable to bioaccumulate, and other substances or groups of substances the levels of which are of concern. Plastics may contain hazardous substances listed in the directive (so-called priority substances), i.e., di(2-ethylhexyl) phthalate (DEHP), nonylphenol, octylphenol, and polycyclic aromatic hydrocarbons (PAHs), the release of which into the environment should be controlled and significantly reduced. Therefore, according to the WFD, pollution caused by discharges, emissions, or losses of priority hazardous substances should be stopped or phased out. In this context, specific measures should be taken to address water pollution by these substances, taking into account all significant sources and identifying economically effective means of and their control (Farmer, 2020). In this area, it is also important to note that Directive 2020/2184, following the European Citizens' Initiative on the Right to Water, the so-called Right2Water initiative, introduces more stringent drinking water quality standards to ensure a higher level of human health protection from pollutants, i.e., endocrine disrupters and microplastics, and to take measures to ensure even cleaner tap water for all – and less need for plastic bottles. Such substances include vinyl chloride, which is listed in Part B of Annex I to Directive 2020/2184 (see Zębek, 2023c).

An important issue for the protection of inland waters against pollution is wastewater treatment, regulated in the EU mainly in Directive 91/271/EEC.

This directive aims to protect the environment in the EU from the adverse effects of urban wastewater, but only in relation to the eutrophication process of water. This directive covers the collection, treatment, and discharge of urban wastewater and the treatment and discharge of waste water from certain industrial sectors. In this respect, Member States should ensure that wastewater treatment plants are properly operated to ensure sufficient efficiency and capacity under all normal weather conditions, and take measures to limit pollution of waters into which wastewater is discharged. Industrial sewage discharged to water should be subject to general principles or regulations and/or specific permits. It is necessary to monitor sewage treatment plants and the waters into which sewage is discharged in order to ensure the protection of the environment from the adverse effects of sewage discharges. Therefore the purpose of Directive 91/271/EEC was to protect the environment from the negative effects of wastewater discharges. In this respect, Member States were obliged to ensure that all agglomerations were equipped with urban wastewater collection systems by 31 December 2000 at the latest for agglomerations with a population equivalent of more than 15,000, and by 31 December 2005 at the latest for agglomerations with a population equivalent of 2,000 to 15,000 (Article 3). Prior to discharge into water bodies, wastewater should be adequately treated by any process and/or disposal system which will enable the receiving waters to achieve the relevant quality objectives and meet the other requirements of this directive and other relevant Community Directives. Discharges from urban wastewater treatment plants meet the relevant requirements of Section B of Annex I of the directive. In addition, the directive provides that water authorities shall monitor discharges from urban wastewater treatment plants to verify compliance with these requirements (Article 15). As part of the legislative process and after many amendments to Directive 91/271/EEC proposed by the European Commission, the European Parliament voted on a new directive. It still has to be adopted by the Council and published in the *Official Journal of the European Union*. The preamble to the new directive emphasises the need to consider water resources in three dimensions: social, economic, and environmental, and to adopt the 'One Health' principle proposed by the World Health Organization (WHO). Article 1 significantly broadens the purpose of collecting, treating, and discharging wastewater. In addition to protecting the environment and human health, it emphasises the need to progressively reduce greenhouse gas emissions, improve the energy balance, ensure universal access to sanitation, increase transparency in the urban wastewater sector, regularly monitor urban wastewater quality parameters relevant to public health, and implement the 'polluter-pays' principle. These changes will undoubtedly be challenging for wastewater treatment plant operators, local governments, and public administrations (EC, 2022b; Ramm, 2024).

Responsibility for the protection of water is also important for the protection of this environmental resource, which is regulated in the EU by Directive 2004/35/EC. The preamble to this directive points out that pollution of the environment, including water, affects health risks and that the loss of biodiversity has accelerated drastically in recent decades. Preventive and remedial measures should therefore be taken to reduce environmental damage in accordance with the 'polluter-pays' principle associated with the principle of sustainable development. The essence of this directive should therefore be that the operator whose activity has caused environmental damage or the threat of such damage should be financially liable in order to force operators to take measures and improve practices to minimise the risk of environmental damage, thus reducing their financial liabilities. The directive defines water damage as any damage that significantly affects the ecological, chemical, and/or quantitative status and/or ecological potential. Remedying water-related environmental damage is achieved by restoring the environment to its original condition through primary, complementary, and compensatory remedial actions. For this purpose, the following actions have been considered such as primary, complementary, and compensatory remediation. Primary remediation depends on any remedial action that restores damaged natural resources and/or diminished services to their original condition. Compensatory remediation is any remedial action to natural resources and/or services that compensates for the fact that primary remediation does not result in the full restoration of damaged natural resources and/or services. However, compensatory mitigation is any action taken to compensate for the temporary decline in natural resources and/or services that occurs between the time of damage and the time when primary mitigation takes full effect (Article 2). In addition, adequate remedial measures should be implemented using the best available technology based on the following criteria: (a) the impact of each option on public health and safety, (b) the extent to which each option will prevent future damage and avoid additional damage as a result of implementing the option, (c) the ranking of the benefits of each option for each element of the natural resource, (d) the length of time it will be effective in restoring the environment to its pre-damage state, and (e) the extent to which each option achieves restoration of the site of environmental damage (Bergkamp & Goldsmith, 2013; Zębek, 2018).

## Summary

The legal protection of inland waters in the EU is closely linked to the implementation of an environmental protection policy, currently the 8th EAP. The need to establish such a policy is included in the provisions of the Treaty, and the idea itself reflects the SDGs. The policy also includes the proper management and protection of water resources in the context of achieving zero pollution as one of its priority objectives. The EU's milestone initiatives in this area are the establishment of the 'Pathway to a Healthy Planet for All: Towards Zero Pollution for Air, Water and Soil', which establishes a new approach to the problem of pollution by introducing a new hierarchy of actions for the elimination of pollution, based on the precautionary principle, and the principles of preventive action, 'repair of damage at source', and 'polluter pays'. This is a very good change because pollution should be prevented at source and then eliminated. Such an approach should prevent significant pollution of surface waters, particularly from industrial discharges. With regard to the protection of water, especially water used for drinking, the European Citizens' Initiative Right2Water was important and led to the adoption of a new Directive 2020/2184, which strengthens the legal protection of such water and, in a sense, guarantees its adequate quantity and quality. In the era of climate change and the resulting water deficits in the EU, it is very important to use water rationally and to reduce losses.

The key legal act establishing the framework for the management and protection of inland waters is the WFD. However, many EC reports show that there are difficulties in implementing its provisions in the Member States, as discussed in the 'Introduction'. Support should therefore be considered, particularly in the form of technical and economic instruments. An important issue in the EU is the proper treatment of wastewater and the elimination of discharges of priority substances, which requires legislative changes, some of which have already been made. Amendments have been made to the Waste Water Treatment Directive in this respect. In this context, more legislative activities are focused on the revision of the list of priority substances, where the protection of water from microplastic pollution is an important aspect. In this way, the principles of management and protection of these waters have been comprehensively incorporated, including the issue of wastewater discharges and liability for damage to the aquatic environment caused by pollution. The aim is to achieve environmental objectives and improve the quality of inland waters in the EU under sustainability conditions.

## References

Backhaus, T. (2023). Commentary on the EU Commission's proposal for amending the Water Framework Directive, the Groundwater Directive, and the Directive on Environmental Quality Standards. *Environmental Sciences Europe* 35 (22): 1–7. https://doi.org/10.1186/s12302-023-00726-3

Bergkamp, L., and B. Goldsmith, eds. (2013). *The EU Environmental Liability Directive: A Commentary*. Oxford.

Boeuf, B., O. Fritsch, and J. Martin-Ortega (2016). Undermining European environmental policy goals? The EU water framework directive and the politics of exemptions. *Water* 8:1–15. https://doi.org/10.3390/w8090388

Bouleau, G., and D. Pont (2015). Did you say reference conditions? Ecological and socioeconomic perspectives on the European Water Framework Directive. *Environmental Science and Policy* 47:32–41. https://doi.org/10.1016/j.envsci.2014.10.012

Carpenter, E.J., and K.L. Smith (1972). Plastics on the Sargasso Sea surface. *Science* 175:1240–1241.

Ciechanowicz-McLean, J. (2013). Ramowa Dyrektywa Wodna a ochrona środowiska morskiego. *Prawo Morskie* 29:107–115.

Dusaucy, J., D. Gateuille, Y. Perrette, and E. Naffrechoux (2021). Microplastic pollution of worldwide lakes. *Environmental Pollution* 284:117075. https://doi.org/10.1016/j.envpol.2021.117075

EC (2009). Staff Working Document accompanying the Report from the Commission to the European Parliament and the Council in accordance with article 18.3 of the Water Framework Directive 2222/60/EC on programmes for monitoring of water status, sec. 415, Brussels.
EC (2010). Common Implementation Strategy for the Water Framework Directive (2000/60/EC), Guidance Document No. 25, Guidance on Chemical Monitoring of Sediment and Biota under the Water Framework Directive, Luxembourg.
EC (2014). Common Implementation Strategy for the Water Framework Directive (2000/60/EC), Guidance Document No. 32 on Biota Monitoring (The Implementation of Eqsbiota) under the Water Framework Directive, Luxembourg.
EC (2019a). Commission Staff Working Document Fitness Check of the Water Framework Directive, Groundwater Directive, Environmental Quality Standards Directive and Floods Directive, SWD (2019) 439 final, Brussels.
EC (2019b). Communication from the Commission to the European Parliament, the European Council, the Council, the European Economic and Social Committee and the Committee of the Regions, The European Green Deal, COM (2019) 640 final, Brussels.
EC (2020a). Commission Staff Working Document Evaluation of the Directive 2008/99/EC of the European Parliament and of the Council of 19 November 2008 on the protection of the environment through criminal law (Environmental Crime Directive), SWD (2020) 259 final, Brussels.
EC (2020b). Communication from the commission to the European Parliament, the Council, the European Economic and Social Committee and the Committee of the Regions: EU Biodiversity Strategy for 2030, COM/2020/380 final, Brussels.
EC (2021). Communication from the Commission to the European Parliament, the Council, the European Economic and Social Committee and the Committee of the Regions, Pathway to a Healthy Planet for All EU Action Plan: Towards Zero Pollution for Air. Water and Soil, COM (2021) 400 final, Brussels.
EC (2022a). Proposal for a directive amending the water framework directive, the groundwater directive and the environmental quality standards directive, COM (2022) 540 final, Brussels.
EC (2022b). Commission Staff Working Document Impact Assessment Accompanying the Document Proposal for a Directive of the European Parliament and of the Council Concerning Urban Wastewater Treatment, SWD/2022/541 final, Brussels.
EC (2023). Proposal for Commission Regulation (EU) amending Annex XVII to Regulation (EC) No. 1907/2006 of the European Parliament and of the Council concerning the Registration, Evaluation, Authorisation and Restriction of Chemicals (REACH) as regards synthetic polymer microparticles, C (2023) 6419, Brussels.
EEA. (2020, August 20). *EU Biodiversity Strategy for 2030*. https://www.eea.europa.eu/policy-documents/eu-biodiversity-strategy-for-2030-1
European Parliament (2021, August 20). *8th EU Environment Action Programme Parliament Europejski*, July 2021. https://www.europarl.europa.eu/RegData/etudes/ATAG/2021/690692/EPRS_ATA(2021)690692_PL.pdf
Farmer, A. (2012). *Manual of European Environmental Policy*. Routledge.
Farmer, A. (2020). *Protecting Europe's soils, protecting Europe's water bodies?EU water law and its ability to support soil protection*. Briefing for iSQAPER by IEEP.
Giakoumis, T., and N. Voulvoulis (2018). Progress with monitoring and assessment in the WFD implementation in five European River Basins: significant differences but similar problems. *European Journal of Environmental Sciences* 8 (1):44–50. https://doi.org/10.14712/23361964.2018.7
Hermoso, V., S.B. Carvalho, S. Giakoumi, D. Goldsborough, S. Katsanevakis, S. Leontiou, V. Markantonatou, B. Rumes, I.N. Vogiatzakis, and K.L. Yates (2022). Biodiversity Strategy for 2030: Opportunities and challenges on the path towards biodiversity recovery. *Environmental Science & Policy* 127:263–271. https://doi.org/10.1016/j.envsci.2021.10.028

Horton, A.A., A. Walton, D.J. Spurgeon, E. Lahive, and C. Svendsen (2017). Microplastics in freshwater and terrestrial environments: evaluating the current understanding to identify the knowledge gaps and future research priorities. *Science Total Environment* 586:127–141. https://doi.org/10.1016/j.scitotenv.2017.01.190

Karbalaei, S., P. Hanachi, T.R. Walker, and M. Cole (2018). Occurrence, sources, human health impacts and mitigation of microplastic pollution. *Environmental Science and Pollution Research* 25:30382517. https://doi.org/10.1007/s11356-018-3508-7

Kentin, E. (2018). Restricting microplastics in the European Union: Process and criteria under REACH. *The European Physical Journal Plus* 133:425. https://doi.org/10.1140/epjp/i2018-12228-2

Pedersen, A.B., T. Hickmann, O. Renn, N. Eckert, K. Jax, R. Lepenies, H.-Y. Liu, J. Lyytimaki, S. Reis, and G. Rusch (2023). SDGs at the halfway point: How the 17 global goals address risks and wicked problems. *Ambio* 52:680.

Pindaru, L.C., A. Nita, I.M. Niculae, S. Manolache, and L. Rozylowicz (2023). More streamlined and targeted. A comparative analysis of the 7th and 8th Environment Action Programmes guiding European environmental policy. *Heliyon* 9 (9):e19212. https://doi.org/10.1016/j.heliyon.2023.e19212

Ramm, K. (2024). Evolution of the EU policy approach to urban wastewater treatment. *Gaz, Woda i Technika Sanitarna* 2024: 13–18. https://doi.org/10.15199/17.2024.6.3

Vlachopoulou, M., D. Coughlin, D.M. Forrow, S. Kirk, P. Logan, and N. Voulvoulis (2014). The potential of using the Ecosystem Approach in the implementation of the EU Water Framework Directive. *Science of the Total Environment* 470–471:684–694. https://doi.org/10.1016/j.scitotenv.2013.09.072

Voulvoulis, N., K.D. Arpon, and T. Giakoumis (2017). The EU Water Framework Directive: From great expectations to problems with implementation. *Science of The Total Environment* 575:358–366. https://doi.org/10.1016/j.scitotenv.2016.09.228

Zębek, E. (2017). Międzynarodowe i krajowe podstawy prawne i bioindykatory (glony) oceny stanu jakości wód powierzchniowych. In *Odpowiedzialność za środowisko w ujęciu normatywnym*, edited by E. Zębek and Hejbudzki, 119–138. UWM Press.

Zębek, E. (2018). Legal protection of waters in the context of human rights. *Themis Polska Nova* 1 (13):14–24.

Zębek, E., and A. Napiórkowska-Krzebietke (2019). Rozwój przepisów prawnych w zakresie bioindykacji środowiskowej a stan jakości wód jeziorowych. *Studia Prawnoustrojowe* 43:375–393. https://doi.org/10.31648/sp.4616

Zębek, E. (2022). Legal solutions of lake monitoring systems in Poland in compliance with the Water Framework Directive. *Review of European and Comparative Law* 48 (2):173–201. https://doi.org/10.31743/recl.13419

Zębek, E. (2023a). European Union environmental policy toward society. In *Family, Law and Society: from Roman Law to the Present Day*, edited by E. Dawidziuk, P. Panero and A. Tarwacka, 499–517. Ius, Lex et Res Publica Studies in Law, Philosophy and Political Cultures, 28. Peter Lang. https://doi.org/10.3726/b21356

Zębek, E. (2023b). Water deficits in the European Union as a consequence of climate change – legal and ecological aspects. *Studia Prawnoustrojowe* 59:389–407. https://doi.org/10.31648/sp.8633

Zębek, E. (2023c). Prawna ochrona wód powierzchniowych przed zanieczyszczeniem mikroplastikiem w UE i w Polsce. *Gdańskie Studia Prawnicze* 4 (61):159–177. https://doi.org/10.26881/gsp.2023.4.09

# 2 Legal protection of surface waters in Czechia

*Veronika Tomoszková, Elżbieta Zębek, and Dominik Zidek*

## Introduction

This study analyses the legal system of surface water protection, including the legal instruments and administrative and criminal liability of entities which use surface water resources in Czechia. The Water Framework Directive (WFD) provisions were implemented into the Water Law of 2001 – CzWL (*Zákon č. 254/2001 Sb., o vodách a o změně některých zákonů (vodní zákon)*). The objective of the law – apart from other things – is to protect waters as an endangered and irreplaceable element of the environment and natural resource, to establish the conditions for the frugal use of water resources, and to maintain and improve their quality, which is also intended to contribute to the protection of aquatic and land ecosystems that rely on them (§ 1 sect. 1 CzWL).

To add the constitutional context to the protection of waters in Czechia, we must mention Art. 7 of the Constitution of the Czech Republic (Constitutional Act No. 1/1993 Coll., *ústavní zákon č. 1/1993 Sb., Ústava České republiky*), according to which 'the State shall ensure the sustainable use of natural resources and the protection of natural wealth'. The case law of the Constitutional Court indicates a tendency towards an extensive interpretation of the concepts of natural resources and natural wealth (Vomáčka & Tkáčiková, 2022), not excluding the protection of water resources. In the Czech legal order, environmental protection, including the protection of water, is conceived as a shared responsibility of all actors, including the state and its organs, territorial self-government units (regional and local), and private persons (natural and legal). Responsibility of private persons is based on Art. 35 par. 3 of the Charter of Fundamental Rights and Freedoms (Act No. 2/1993 Coll., *Usnesení předsednictva České národní rady o vyhlášení LISTINY ZÁKLADNÍCH PRÁV A SVOBOD jako součásti ústavního pořádku České republiky*), which stipulates that '[i]n exercising his or her rights, no one may endanger or damage the environment, natural resources, biodiversity and cultural monuments beyond the extent prescribed by law'. The purpose and intent of this provision are not to prohibit across the board

DOI: 10.4324/9781003586913-3

all potentially hazardous activities to the environment but rather to legitimise legal measures that restrict or impose conditions on the exercise of various rights on the grounds of environmental protection. Without such restrictions, it would be left entirely to the discretion of the individual to determine how far he or she would take the environment into account in exercising his or her rights (Vomáčka et al., 2021; Židek, 2021).

Based on § 3 sect. 1 CzWL surface water and groundwater are not subject to ownership and are not part of or appurtenant to the land on or under which they occur. Water becomes property only when it is withdrawn from its source (§ 3 sect. 2 CzWL). From the point of view of their nature, surface, and groundwater are considered to be *res nullius*, although some studies also classify water as a thing outside commerce (*res extra commercium*; Huneš & Tomoszková et al., 2022).

In the last four years the Czech water law underwent significant changes due to adoption of the new Building Act (Act No. 283/2021 Coll., *zákon č. 283/2021 Sb., stavební zákon*) and of the new Act on Single Environmental Statement (Act No. 148/2023 Coll., *zákon č. 148/2023 Sb., o jednotném environmentálním stanovisku*). The experience with one of the worst environmental accidents in the Czech Republic in the last ten years, which took place in September 2020 on the river Bečva, where 40 kilometres of the river near Přerov were contaminated by cyanide from an unknown source clearly demonstrated the shortcomings of the Czech legislation hindering the effective handling of accidents. The excessive complexity and, at the same time, the lack of coherence of the Czech law also complicate the practical application of liability for environmental damage (Česká společnost pro právo životního prostředí, 2021). In June 2024, the Czech Parliament approved the so-called emergency amendment (*havarijní novela*) of the CzWL, which aims to rectify the shortcomings of the CzWL in respect to handling accidents (Act No. 182/2024 Coll., *zákon č. 182/2024 Sb., kterým se mění zákon č. 254/2001 Sb., o vodách a o změně některých zákonů (vodní zákon), ve znění pozdějších předpisů, zákon č. 114/1992, o ochraně přírody a krajiny, ve znění pozdějších předpisů, a zákon č. 465/2023 Sb., kterým se mění zákon č. 416/2009 Sb., o urychlení výstavby dopravní, vodní a energetické infrastruktury a infrastruktury elektronických komunikací (liniový zákon), ve znění pozdějších předpisů, a další související zákony*).

## Legal basis of surface water management and protection

The protection objectives for water as an element of the environment implementing the objectives of the WFD are included in § 23a CzWL: (a) preventing deterioration of all the water bodies; (b) ensuring protection, improvement, and restoration of all the water bodies and achieving its good status; (c) ensuring protection, improvement, and achievement of good ecological potential and good chemical status of all artificial and heavily impacted water bodies; and (d) reducing their pollution with hazardous substances (*závadné látky*), as well as discontinuation of their emissions or gradual elimination of discharges and leakages of dangerous hazardous substances (*nebezpečné závadné látky*).

The legislator established the rules for water management in § 5–20 CzWL. Water management (*nakládání s vodami*) as a central term is defined in § 2 sect. 9 CzWL as raising of surface or

> groundwater by means of water works, the exploitation of its energy potential, its use for navigation or floating of timber, the rearing of fish or waterfowl, its abstraction, the discharge of wastewater into it and other ways in which its properties can be exploited or its quantity, flow, occurrence or quality influenced.

From a theoretical point of view, the legal regime of water management (*nakládání s vodami*) represents an example of public use (*veřejné užívání*), which can be divided into (1) general use (*obecné užívání*) and (2) special use (*zvláštní užívání*). In theory, general use means that anyone can use a certain thing or a natural resource dedicated for public use for her or his own personal need, at her or his own risk, usually without any permission or authorisation needed and usually free of charge (Staša, 2012; Kopecký, 2023). As opposed to that, special use requires authorisation or permission based on which it allows only a specific person (addressee of the authorisation or permission) to use a thing or natural resource dedicated for public use beyond general use (beyond personal need) and in a way and according to conditions specified in the authorisation. Special use is usually associated with a specific fee (Staša, 2012; Kopecký, 2023).

The legal regime of general use (*obecné užívání*), i.e., anyone can use a natural resource (water) for her or his own personal need, at her or his own risk, without any permission or authorisation and free of charge, only applies to surface water. Based on the Czech law, groundwater may only be used and managed within the legal regime of special use of water (*zvláštní užívání*) (Huneš & Tomoszková et al., 2022).

Before listing specific rules of surface water management, we must mention general obligations for water management set in § 5 CzWL, which serves for interpretation purposes and includes basic obligations (sect. 1), obligations regarding water usage for production purposes (sect. 2), and for building purposes (sect. 3) (Horáček et al., 2015).

According to § 5 sect. 1 CzWL anyone who handles surface water or groundwater is obliged to take care to protect it and to ensure its economical and efficient use in accordance with the CzWL and to ensure that its energy potential is not devalued and that other public interests protected by special legislation are not violated.

Those who handle surface or groundwater for production (business) purposes shall, in order to comply with the obligations under sect. 1, make effective adjustments in the production process leading to the economical use of water resources and taking into account the best available technology (§ 5 sect. 2 CzWL).

What concerns building purposes (constructing new buildings, carrying out their changes, or changes of their usage), the builder is obliged, according to the nature and purpose of the use of these buildings, to provide them with water supply and sewage disposal. For the sewage disposal, § 5 sect. 3 CzWL sets a hierarchy of wastewater disposal methods: (1) through a sewerage system designed for this purpose, (2) if no sewerage system is available on site, the wastewater shall be disposed of by direct treatment with subsequent discharge to surface or groundwater, and (3) if methods (1) and (2) are not possible, wastewater may be accumulated in an impermeable sump (cesspit) with subsequent disposal of the accumulated water at a facility approved for its disposal. Furthermore, the builder is obliged to ensure the limitation of the surface water runoff resulting from the impact of atmospheric precipitation on these buildings by accumulation and subsequent use, or seepage on the land, by evaporation, or, if none of these methods of limiting the runoff of rainwater is possible or sufficient, by their retention and controlled drainage or a combination of these methods. The aforementioned obligations concerning wastewater as well as rainwater need to be met; otherwise, building, alteration of a building before its completion, use of a building, decision on a supplementary building permit, or decision on a change in the use of a building shall not be permitted.

General use (*obecné užívání*) of surface water is regulated in § 6 and 7 CzWL. Based on § 6 sect. 1, anyone may, at his or her own risk, withdraw surface water or otherwise dispose of it for his or her own personal use without any permission or consent of the water authority unless special technical facilities for water usage are required. Permission or consent from the water authority is also not required for the interception of surface water by simple devices on individual plots and buildings or for the alteration of the natural flow of water in order to protect it from the harmful effects of such water (§ 6 sect. 2 CzWL).

The use of surface waters for navigation and the abstraction of water required for the operation of vessels also does not require a permit from the water authority (§ 7 sect. 1 CzWL) and is considered to be part of the general use legal regime. The CzWL and Act No. 114/1995 Coll. on Inland Navigation (*zákon č. 114/1995 Sb., o vnitrozemské plavbě*) set further rules for protecting surface water while it is used for navigation purposes.

Based on § 6 sect. 3 CzWL, general use of surface water cannot result in its quality deterioration or a decrease in its protection level, upsetting the natural environment, deterioration of the drainage conditions, destroy the banks, water supply lines or other water objects, fish breeding, or violate the laws and regulations or infringe on other persons' rights. The water authority may modify, restrict, or prohibit the general use of surface water without compensation if the public interest so requires, in particular if the obligations under § 6 sect. 3 CzWL are violated or for reasons of personal safety (§ 6 sect. 4 CzWL).

Special use (*zvláštní užívání*) of water requires some type of administrative act (permit, approval, binding opinion, or opinion) issued by the water authority and is usually connected with the obligation to pay a charge (charges are regulated in § 88–90i CzWL).

In the Czech water law, the following types of legal instruments may be distinguished:

1 informative – a monitoring system for water quantity and quality, water management registers and relevant public administration information systems,
2 planning – national and regional water management plans and programmes,
3 administrative – defining general responsibilities and prohibitions for water users and a system of permits and other administrative acts,
4 economic – system of charges for water management,
5 corrective – remedial measures in the event of water pollution in emergency situations,
6 liability and sanctions, and
7 special – concerning floods and drought (Sobotka, 2010; Müllerová, 2015; Strnad et al., 2015).

In the following text, we will introduce the Czech water law in greater detail.

The system of water protection instruments allows for the regulation of activities that may lead to the deterioration of water quality, threaten the abundance and quantity of water, or disturb water-related ecosystems. Effective measures to protect surface waters can only be implemented based on knowledge of water conditions and the nature of activities that affect them (Sobotka, 2010).

In order to gather information, a system of monitoring and assessment of the status of surface and groundwater is established (§ 21, 22 CzWL). Within this system, information on the status and quantity of water is collected, a water balance (*vodní balance*) is kept, and relevant records with data within the scope of § 21 sect. 2 lit. c) CzWL are maintained and updated. Relevant data from the records, including statistical and cartographical data are forwarded to public administration information systems, which are operated by the river basin managers (*správci povodí*): the Czech Hydrometeorological Institute (*Český hydrometeorologický ústav*) and the T. G. Masaryk Research Institute of Water Management (*Výzkumný ústav vodohospodářský T. G. Masaryka, v.v.i.*) (Horáček et al., 2015).

Based on § 19 CzWL, the water authorities are obliged to maintain the register of all administrative acts issued by them and upload these documents to the public administration information system operated either by the Ministry of Agriculture (§ 22 sect. 3 CzWL) or by the Ministry of Environment (§ 22 sect. 4 CzWL). The CzWL (§ 20) also regulates what information shall be registered in the real estate cadastre (*katastr nemovitostí*) – this applies to dams, weirs, structures erected in river canals for the purpose of navigation in watercourses or on their banks, structures aiding in using hydro power, and construction of wastewater treatment plants if they are connected to the ground with a solid foundation (§ 20 sect. 1 CzWL).

The Ministry of Agriculture maintains the public administration information system (§ 22 sect. 3 CzWL), which contains a register of data concerning water courses and their drainage basins, hydrogeological areas and water bodies, surface water intake, and wastewater discharge and its accumulation in water tanks. The Ministry of Environment maintains a public administration information system (§ 22 sect. 4 CzWL) concerning data on the registration of the quality and quantity of water, the condition of water bodies and their ecological condition, protected areas of natural accumulation of water, areas susceptible to pollution and flood plains.

With the so-called emergency amendment to the CzWL (Act No. 182/2024 Coll.), which was approved in the Parliament of the Czech Republic in June 2024, a new register of discharges from pollution sources into surface waters shall be established to serve as an information system for investigating the causes of accidents and for collecting information on potential sources of threats to water quality (§ 19a CzWL). The well-publicised accident on the Bečva River, which took place in September 2020, showed that the basic shortcoming of the Czech legislation was the lack of an overview of discharges from which pollution may leak into surface waters.

In line with the European Union (EU) law, the CzWL also contains legal provisions for water protection planning in the form of river basin management plans (international river basin management plans, national river basin management plans, and sub-basin management plans) and flood risk management plans. The CzWL lays down the procedure for drawing up both types of plans, including public participation (§ 25 CzWL). Further details are laid down in Decree No. 50/2023 Coll. on River Basin Management Plans and Flood Management Plans (*vyhláška č. 50/2023 Sb., o plánech povodí a plánech pro zvládání povodňových rizik*). River basin management plans and flood risk management plans are issued in the form of measures of a general nature (*opatření obecné povahy*, a specific mixed type of administrative act). Both categories of plans form a background document for spatial planning, water law proceedings, and administrative proceedings under the Building Act (Act No. 283/2021 Coll.). They are reviewed and updated every six years after their approval.

On the basis of the river basin management plans and flood risk management plans, the so-called programmes of measures (§ 26 CzWL) are developed, containing specific measures for achieving the objectives set out in the plans, a timetable for their implementation and a method of financing them (Sobotka, 2010). The CzWL stipulates that the measures set out in the programme of measures must be implemented within three years of the approval of the river basin plans (§ 26 sect. 1). If, in the course of the surface water quality monitoring and assessment pursuant to § 21 CzWL or from other data, it is found that the water protection objectives set for the relevant water body are unlikely to be achieved, § 26 sect. 4 CzWL provides for a further procedure for investigating the causes and taking additional measures within the process of implementing the respective programme of measures.

The range of human activities in which water conditions can be affected is wide, as is the range of persons who carry out these activities (Sobotka, 2010). Obligations for persons whose activities could affect water conditions are laid down either directly by law (CzWL or other) or by an administrative act. The CzWL contains specifics for the so-called water management administrative procedure (*vodoprávní řízení*, § 115 CzWL) and defines the activities for which a specified administrative act must be obtained (§ 8–17 CzWL).

An administrative decision (water permit, *vodoprávní povolení*) is required for water management and other activities that have the most significant impact on water status. Based on § 8 CzWL, the water permit is required for water use outside the legal regime of general use (*obecné užívání*), which does not require special technical devices, e.g., surface water intake, retention, and storage, water use for energy generation and for economic purposes, as well as for wastewater discharge to surface water, pumping polluted groundwater in order to reduce its pollution and for its later discharge to surface waters. It is crucial to note that a valid water permit, which can only be exercised through the use of a waterworks, is a condition for the enforceability of the permit for that waterworks under the Building Act. Moreover, a permit is required to discharge wastewater containing especially dangerous hazardous substances (*zvlášť nebezpečné závadné látky*) or priority dangerous substances (*prioritní nebezpečné látky*) which are harmful to the sewerage system (§ 16 CzWL).

Pursuant to § 11 sect. 2 CzWL a water permit does not create rights to other people's land and buildings, nor does it create a legal obligation for the water authority, watercourse manager (*správce vodního toku*) or waterworks owner to compensate the water permit beneficiaries for the inability to use water in the maximum permitted quantity and to have available water with certain characteristics. Moreover, according to § 12 CzWL, the water authority can – on its own initiative or on request – change the rules of water management laid down in the water permit or rescind such a permit, if necessary

a to achieve the water protection objectives adopted in the river basin management plan (§ 24 and 26 CzWL),
b to implement the programmes of reduction of surface water pollution (§ 34 sect. 2 and § 35 sect. 1 CzWL),
c to comply with the programme of reduction of surface water pollution by dangerous hazardous substances and especially dangerous hazardous substances (§ 39 sect. 3 CzWL), and
d for the supply of drinking water in accordance with the water supply and sewerage development plan pursuant to the Act No. 274/2001 Coll., on Water Supply and Sewerage, *zákon č. 274/2001 Sb., o vodovodech a kanalizacích*).

The application for each water permit is accompanied by an opinion of the river basin district manager (§ 54 sect. 4 CzWL), in which the river basin manager (*správce povodí*) assesses the possibility of deterioration of the good status or ecological potential of the water body concerned or the impossibility of achieving good status or good ecological potential of the water body concerned (§ 23a sect. 7 CzWL). In the event that the river basin district manager, in its opinion, indicates the possibility of deterioration or even the impossibility of achieving the set objectives, the applicant may obtain a water permit only if an exemption is granted under the conditions laid down in § 23a sect. 8 CzWL.

In addition to administrative decisions, water authorities also issue

- binding opinion (*závazné stanovisko*), i.e., a binding background document for a decision of another administrative authority – § 149 sect. 1 of the Administrative Procedure Code (*zákon č. 500/2004 Sb., správní řád*);
- measure of a general nature (*opatření obecné povahy*), i.e., a mixed administrative act issued in a specific matter to an unspecified range of addressees, which is not an administrative decision or a legal regulation; or
- statement (§ 17 CzWL, in Czech *vyjádření*) for anyone who intends to carry out activities that may affect the water conditions, energy potential, quality or quantity of surface water (or groundwater) to obtain information from the water authority, after sufficient evidence of the intended project, as to whether his/her project is possible in terms of the interests protected under the CzWL, and under what conditions.

In order to protect the purity (quality) of water, the CzWL contains legal regulations for the two most important sources of pollution: the management of wastewater and pollutants. Both legal regimes are described in more detail in this chapter.

In order to protect the quantity of surface water, the CzWL regulates the so-called minimum residual flow (*minimální zůstatkový průtok*), which is set by the water authority in the water management permit (§ 36 CzWL). In doing so, the water authority takes into account the conditions of the particular watercourse and recreational navigation opportunities and the nature of the water management and bases its decision on measures to achieve the water protection objectives adopted in the river basin management plan and programmes of measures. The water authority shall also determine the place and method of measurement of the minimum residual flow and the frequency of submission of the measurement results to the water authority.

Protection of water quantity and purity cannot be ensured by protecting water alone, as these are influenced not only by the management of water itself but also by the activities that take place in the area where the water is located. Therefore, the CzWL also enshrines the tools of territorial water protection (Sobotka, 2010). Landowners are obliged to take care of the land in such a way that water conditions do not deteriorate. Special protection is provided for the territories of (1) protected areas of natural water accumulation (*chráněné oblasti přirozené akumulace vod – CHOPAV*), (2) areas protected for surface water accumulation (*území chráněná pro akumulaci povrchových vod*), (3) water resource protection zones (*ochranné pásmo vodního zdroje*), and (4) vulnerable areas (*zranitelné oblasti*).

The CzWL also pays special attention to the protection of watercourses (*vodní toky*), which contain most of the surface water and are its most important source (Sobotka, 2010). The total length of watercourses on the territory of the Czech Republic is 102,879 km – of which the major watercourses (*významné vodní toky*), namely rivers *Labe*, *Vltava*, *Ohře*, *Odra*, and *Morava* make-up 16,326 km, and other watercourses referred to as 'small watercourses' (*drobné vodní toky*) make-up the rest of 86,553 km.

Watercourses and their beds are protected from activities that could affect the natural course. To ensure the proper condition and care of watercourses, the CzWL establishes the so-called watercourse management (*správa vodních toků*, § 47 CzWL). This is an activity carried out by special public authorities, the so-called watercourse managers (*správci vodních toků*, § 48 CzWL). The watercourse managers monitor the condition of the watercourse and its bed, carry out the necessary maintenance work and create the conditions for the implementation of the water management authorisations.

In the case of major watercourses (*Labe*, *Vltava*, *Ohře*, *Odra*, and *Morava*), their management is carried out by the river basin managers (*správci povodí*), who, in addition to managing the watercourses, also manage the entire river basin, which includes, among other things, surface water quality monitoring and assessment in the given river basin. The management of small watercourses or their complete sections is carried out by the municipalities through whose territory the small watercourses flow, by natural or legal persons or organisational units of the state to which the small watercourses serve or whose activities they are related to, or by state enterprises on the basis of a designation by the Ministry of Agriculture (§ 48 sect. 2 CzWL).

The economic instruments of water protection consist of a system of charges, including charges for water pollution and charges for the use of water as a natural resource (Sobotka, 2010). The charge system consists, namely, of (1) charges for the discharge of wastewater into surface water, (2) charges for the discharge of wastewater into groundwater, (3) payment for the management of watercourses and river basins, and (4) a charge for the amount of groundwater withdrawn (§ 88–90i CzWL).

The aforementioned instruments are complemented by a system of administrative supervision (*vodoprávní dozor*), liability for environmental damage and by administrative and criminal liability. Further details on the liability regime are provided in this chapter.

### Water authority – structure, responsibilities, and competences

The national administration in Czechia, in accordance with § 104 CzWL, is performed by water authorities (*vodoprávní úřady*) and the Czech Environmental Inspectorate (*Česká inspekce životního prostředí*). The water authorities include

(a) municipal authorities (*obecní úřady*), (b) district authorities in the military districts[1] (*újezdní úřady ve vojenských újezdech*), (c) municipal authorities of municipalities with extended competences (*obecní úřady obcí s rozšířenou působností*), (d) regional authorities (*krajské úřady*), (e) building authorities *(stavební úřady)*, and (f) ministries as central water authorities.

There are four ministries which act as a central water authority: the Ministry of Agriculture, the Ministry of Environment, the Ministry of Transport, and the Ministry of Defence (§ 108 of the CzWL). Other defined bodies established under the provisions of the CzWL are river basin managers (*správci povodí*, § 54 and § 82 CzWL) and watercourse managers (*správci vodních toků*, § 48 and § 83 CzWL).

Article 8 of the Czech Constitution guarantees self-government of territorial self-government units on the local level (6,258 municipalities)[2] and the regional level (14 regions). In Czechia, public administration operates as a mixed model (Průcha, 2008), which means that the state administration is performed by the state and its bodies (ministries and other central and regional state institutions) but also by offices of regional and municipal self-governing units upon which the state delegates certain state administration tasks (Kopecký, 2023; Sládeček, 2019).

The Czech self-governing regions (*kraje*), which were established in 2000, are equivalent to the Polish Voivodship. Since coming into effect on 1 January 2003, the regional offices (*krajské úřady*) of the self-governing regions took over part of the state administration tasks of the former district offices (*okresní úřady*), which served as state administration bodies on district level and which were abolished in 2002 (Kopecký, 2023; Sládeček, 2019). The other part of the state administration tasks of the abolished district offices was delegated to municipal offices (*obecní úřady*) of the self-governing municipalities.

Depending on the scope of the exercise of delegated state administration, we distinguish three categories of municipalities: (1) municipalities with basic municipal authority, which they perform only for their own territory; (2) municipalities with delegated municipal authority (*obce s pověřeným obecním úřadem*), which they perform not only for their own territory but also for other, usually neighbouring municipalities; and (3) municipalities with extended competence (*obce s rozšířenou působností*). Most of the state administration tasks of the former district offices (*okresní* úřady) were delegated to municipalities with extended competences. All 6,258 municipalities exercise basic municipal authority. Beyond that, there are 388 municipalities with delegated municipal authority, and out of these, 205 municipalities exercise state administration to the widest extent (municipalities with extended competence).

If we look more closely at the division of competencies on the central level, these are pursuant to § 15 and § 19 of the Act No. 2/1969 Coll. (the so-called Competence Act, *kompeteční zákon, zákon č. 2/1969 Sb.*) shared by the already mentioned four ministries (Strnad et al., 2015). Most of the competencies are shared by the Ministry of Agriculture and the Ministry of Environment. The CzWL defines the competencies of the Ministry of

Environment as an exhaustive list (§ 108 sect. 3 CzWL), while, based on the residual clause in § 108 sect. 1 CzWL, the **Ministry of Agriculture** exercises the powers of the central water authority unless the CzWL provides otherwise (Strnad et al., 2015). At the same time, some of the competencies are shared and based on the necessary cooperation of these two ministries (e.g., establishment and maintenance of the information system pursuant to § 19 and § 22 CzWL or monitoring and assessment of the status of surface water and groundwater pursuant to § 21 and § 22 CzWL).

Pursuant to the exhaustive list of competencies provided in § 108 sect. 3 CzWL, the **Ministry of Environment** exercises the competence of the central water authority in issues related to

a protection of the quantity and quality of surface waters within which the Ministry of Environment:

 1 issues permit for discharges of wastewater to surface water or groundwater;
 2 issues permit for pumping polluted groundwater in order to reduce their pollution and their subsequent discharge to surface waters;
 3 issues permit for discharge to the sewerage system of wastewater containing especially dangerous hazardous substances or priority dangerous substances;
 4 issues binding opinion to activities listed in § 17 sect. 1 lit. (e) and (i) CzWL;
 5 sets the minimum residual flow and imposes the obligation to place a mark or watermark on the water body and the obligation to measure the minimum residual flow regularly and to report the results of the measurements to the water authority and the river basin district manager (§ 36 CzWL);
 6 sets the minimum groundwater level and imposes the obligation to submit a draft abstraction order or, where appropriate, the obligation to measure the groundwater level regularly, the method of measurement, and the obligation to report the results of the measurement to the water authority and the river basin district manager (§ 37 CzWL);
 7 conducts activities for the authorisation of discharging wastewater to surface waters and acceptance of the results of the volume measurement of the discharged water and the level of its pollution (§ 38 CzWL);
 8 prepares the draft implementing regulation which specifies the indicators and the level of acceptable water pollution (§ 31, 32, 33, 35, and 38 CzWL);

b cooperation with regional authorities in delegated competence in matters of protection of the quality and quantity of border waters (§ 107 CzWL);
c quality monitoring and assessment of surface water (as well as groundwater) condition in collaboration with the Ministry of Agriculture (§ 21 and 22 CzWL) and via authorised expert bodies established by the Ministry of Environment (§ 21 CzWL);

d establishment and maintenance of an information system in collaboration with the Ministry of Agriculture (§ 19 and 22 CzWL);
e drawing up and approving water plans and drawing up programmes of measures jointly with the Ministry of Agriculture (§ 24 to 26 CzWL);
f assessment of environmental impacts of water plans (§ 24 and 25 CzWL);
g protected areas of natural water accumulation (chráněné oblasti přirozené akumulace vod – CHOPAV) (§ 28 CzWL);
h protection of water resources (§ 29 and 30 CzWL);
i sensitive areas (citlivé oblasti) (§ 32 CzWL) and vulnerable areas (zranitelné oblasti), except for the review of the action programme (§ 33 CzWL);
j determination of remedial measures for unsatisfactory quality of surface waters used for bathing (§ 34 sect. 2 CzWL);
k designation of surface waters suitable for breeding and reproduction of native species of fish and other aquatic animals jointly with the Ministry of Agriculture (§ 35 CzWL);
l protection of surface waters (as well as groundwater) against the hazardous substances (§ 39 CzWL);
m disposal of accidents (§ 41 CzWL) and elimination of their harmful consequences (§ 42 CzWL);
n flood protection (§ 63–87 CzWL);
o drought and water scarcity management jointly with the Ministry of Agriculture (§ 87a–87m CzWL);
p management of the Czech Hydrometeorological Institute (Český hydrometeorologický ústav – ČHMÚ) (§ 73 CzWL) and of the Czech Environmental Inspectorate (Česká inspekce životního prostředí) (§ 112 CzWL);
q charges – namely, groundwater abstraction charges, jointly with the Ministry of Agriculture (§ 88–88l CzWL) and charges or discharging wastewater to surface waters (as well as to groundwater) (§ 89–89q CzWL);
r management of basic and applied research with respect to the protection of water quantity and quality;
s performance of tasks arising from the relationship with the EU in the field of water protection, reporting on the implementation of the relevant EU directives, and coordinating the adoption and implementation of EU legislation in the field of water together with the Ministry of Agriculture; and
t ensuring coordination of water plans and programmes of measures within the framework of international cooperation on water protection in the Elbe, Danube, and Oder River basin districts.

On regional and local levels, most of the competencies of water authorities are divided between regions and municipalities with extended competence and are exercised by these self-governing units as delegated state administration (Strnad et al., 2015). The CzWL defines the competencies of regional offices in the form of an exhaustive list (§ 107 CzWL), while the competencies of **municipal offices in municipalities with extended competence** are regulated in the form of residual clause of § 106 sect. 1 CzWL, which states, 'The competence that belongs to water authorities is exercised by municipal authorities of municipalities with extended competence, unless the law entrusts it to other authorities'.

Based on the CzWL, the **regional offices** (*krajské úřady*) as water authorities are competent to provide the following:

1 Authorisation of water management in matters of boundary waters (issuing permits as well as binding opinions based on § 17 or 104 sect. 3 CzWL)
2 Authorisation of specific types of water management pursuant to § 8 or § 16 CzWL:

   a The rising and storage of surface water in reservoirs with a total volume exceeding 1,000,000 m$^3$ or with a height of rise exceeding 10 m from the bottom of the foundation outlet (§ 8 sect. 1 lit. a) no. 2 CzWL)
   b The use of water for the exploitation of its energy potential on significant watercourses where the regional office reserves this competence for itself (§ 8 sect. 1 lit. a) no. 3 CzWL)
   c The wastewater discharge into surface waters from pollution sources of 10,000 equivalent population or more (§ 8 sect. 1 lit. c) CzWL)
   d The pumping of polluted groundwater to reduce its pollution and its subsequent discharge into such waters or, where appropriate, into surface waters (§ 8 sect. 1 lit. e) CzWL)
   e The discharge of wastewater from uranium ore mining and processing and nuclear power stations and wastewater containing especially dangerous hazardous substances or dangerous hazardous substances as defined in Annex 1 into surface waters and the discharge of wastewater containing an especially dangerous hazardous substance or priority dangerous substance into sewers (§ 16 CzWL), except where equipment of sufficient efficiency is installed in accordance with § 16 sect. 5 CzWL

3 Granting exemptions pursuant to § 23a sect. 8 CzWL and exemptions for the use of harmful substances pursuant to § 38 sect. 7 CzWL
4 Decision-making which regards imposing measures in public interest listed in the programmes of measures (§ 26 sect. 4 CzWL)
5 Decision-making which regards measures in emergency situations, in particular in the event of water shortages and accidents, in cases that go beyond the administrative area of the municipality with extended competence or the scope of competence of the municipal authority of the municipality with extended competence
6 Decision-making in various cases of doubt (*rozhodování v pochybnostech*) (e.g., whether surface water or groundwater is involved, what water management is involved or whether wastewater is involved, whether a watercourse is a watercourse as referred to in § 43 CzWL)
7 Issuing a measure of a general nature (*opatření obecné povahy*), in which the regional office, based on the proposal of the river basin manager

(*správce povodí*), determines the area of floodplains of significant watercourses and their active zones and imposes restrictive conditions in the respective floodplain (§ 66 and 67 CzWL)

8 Issuing a measure of a general nature, in which the regional office determines protection zones for water source, for which the regional office is competent to authorise the water management (§ 30 CzWL)
9 Approving handling schedule (*manipulační řád*) for waterworks for which it authorises the water management, including emergency handling thereof, and comprehensive handling schedule (*komplexní manipulační řád*), where the competence of regional office includes authorising the water management for at least one of the waterworks to which the comprehensive handling schedule will apply

Besides acting as decision-making authorities, water authorities take part in various administrative procedures to deliver their opinions (*stanoviska*) or binding opinions (*závazná stanoviska*) from the position of the so-called concerned authorities (*dotčené orgány*) (Strnad et al., 2015).

As the CzWL prescribes, water authorities take part in the spatial planning process and deliver their opinions on spatial planning documents, including national policy documents, as well as planning documentation which bindingly determines the functional use of the territory. Both the Ministry of Environment and the Ministry of Agriculture submit opinions on the Policy of Spatial Development (*politika územního* rozvoje) and Spatial Development Plan (*územní rozvojový plán*), which are both elaborated for the whole territory of Czechia, opinions on the Principles of Spatial Development (elaborated for the territory of each region, *zásady územního rozvoje*) and to the zoning measure on construction closure (*územní opatření o stavební uzávěře*) elaborated by the Ministry of Regional Development or by any of the regions (§ 108 sect. 2 CzWL).

The regional offices as concerned water authorities submit their opinions to the Principles of Spatial Development, to spatial plans and zoning measures of construction closures elaborated for the territory of municipalities with extended competence (§ 107 sect. 1 lit. a) CzWL). Municipal offices of municipalities with extended competence submit opinions on spatial plans and zoning measure on construction closure of other municipalities within the area of their extended competence (§ 106 sect. 2 CzWL).

The aforementioned opinions of the water authorities on spatial planning instruments are not 'binding opinions', but their content is binding. This means that the water authority cannot completely prevent the adoption of (for example) a spatial plan, but it can set binding conditions for land use regulation in water protection.

**Municipal offices** in municipalities with basic municipal authority (so basically all 6,258 Czech municipalities) are competent to regulate, limit, or prohibit general use of surface water (*obecné užívání povrchových vod*) (§ 6 sect. 4 CzWL) except for watercourses delineating state borders. Pursuant to § 115 sect. 4 CzWL any municipality is a party to the administrative proceedings (*účastník správního řízení*) if the water authority's decision may affect water conditions or the environment in its territorial district (Strnad et al., 2015).

Based on the Building Act (Act No. 283/2021 Coll.), which came fully into force on 1 July 2024, the **building offices** (*stavební úřady*) also act as water authorities (§ 104 sect. 2 lit. e) CzWL) and based on the list of competencies in § 107a CzWL, they are responsible for state administration concerning the waterworks (*vodní díla*), including issuing building permits for a waterwork, delimiting protection zone of a waterwork, authorising technical safety inspections and conducting water supervision over waterworks.

In this context, it should be stressed that Czech water law distinguishes between decisions to permit the construction of a waterwork and decisions to permit constructions that are not waterworks but may affect water conditions. Waterworks are defined in § 55 CzWL as structures serving to raise and retain water, artificially direct the flow regime of surface water, protect and use water, manage water, protect against the harmful effects of water, regulate water conditions, or fulfil other purposes of the CzWL. Examples include dams, dykes, reservoirs, weirs, flood protection structures, and wells. Simple installations outside watercourses on land or buildings to capture water and protect it from the harmful effects of surface or groundwater, such as bottomless pits, are not considered waterworks under this CzWL. In case of doubt about whether a structure qualifies as a waterwork, the locally competent water authority shall decide (§ 55 sect. 4 CzWL).

Therefore, permission to construct waterworks requires a permit from the building authority, issued through a procedural process involving both the CzWL and the Building Act. When permitting the construction, alteration, change in use, or removal of waterworks, the application under the Building Act must include the opinion of the river basin authority (see p. 000). This requirement does not apply to individual water supply wells and sewage treatment plants for up to 50 equivalent inhabitants for the needs of individual households. The construction of waterworks must consider the protection of aquatic and water-related ecosystems, ensuring that they do not create barriers to the movement of fish and aquatic animals in both directions of the watercourse and that they can be crossed within the waterway. Exceptions to this rule are made when flood protection or other public interests require it.

The second situation arises when the construction is not a waterwork but it's a structure, installation, or activity that may affect water conditions. In this case, a binding opinion of the water authority is required (§ 17 CzWL), specifically for

a buildings and installations on land containing watercourses or on land adjacent to such areas, where these constructions will affect water conditions;
b the establishment of long-distance pipelines and structures for the underground storage of substances in earth cavities, as well as warehouses, landfills, or tanks, if their operation may significantly endanger the quality of surface or groundwater;

c construction, mineral extraction, or landscaping in floodplains;
d buildings within 15 m from the aerial foot of the protective dam of the watercourse;
e buildings in the protection zones of water sources;
f storage sites for the management of extractive waste or decisions on the obligation to collect and treat contaminated water and leachate pursuant to other legal regulations;
g boreholes for the exploitation of the energy potential of groundwater from which groundwater is not extracted or pumped; unless otherwise provided for in the CzWL, the water authority may, in the procedure for granting such consent, require the applicant to submit a statement from a person with professional competence;
h the storage of carbon dioxide in natural rock structures pursuant to other legislation; or
i geological works associated with land disturbance, intended for subsequent use in groundwater extraction or for drilling to exploit the energy potential of groundwater.

The **Czech Environmental Inspectorate** plays an important role in overseeing the proper application and implementation of water law. It was established in 1991 by the Act No. 282/1991 Coll. (Act on Czech Environmental Inspectorate and Its Competence in Forest Protection, *zákon č. 282/1991 Sb., o České inspekci životního prostředí a její působnosti v ochraně lesa*) as a special supervision body in the area of environmental protection (Strnad et al., 2015). Originally, the Czech Environmental Inspectorate was supposed to exercise the supervision competence only in the area of forest protection, but the scope of competences soon expanded to other areas of environmental protection, including water protection.

The Czech Environmental Inspectorate is subordinate to the Ministry of Environment (§ 1 of the Act No. 282/1991 Coll.). From an administrative point of view, the Czech Environmental Inspectorate does not belong to the water authorities (Strnad et al., 2015), but according to § 112 sect. 1 lit. (d) CzWL, it cooperates with water authorities and State Environmental Fund (*Státní fond životního prostředí*).

In the area of water protection, the competencies of the Czech Environmental Inspectorate are delimited in § 112 CzWL, based on which it *inter alia* exercises the following:

a water law supervision within which it controls how natural or legal persons undertaking business comply with the obligations laid down by the CzWL or imposed on them or by water authorities, namely what concerns the management of surface water (or groundwater); water protection, including protection of water during navigation; accidents endangering the water quality; waterworks intended for the disposal of pollution in wastewater or for its discharge into surface water (or groundwater or sewers);

protection of water conditions and water resources; and bringing the discharge of wastewater into surface or groundwater into compliance with the requirements of the CzWL in cases requiring special attention;

b decision-making on elimination and correction of identified deficiencies, their causes, and their harmful consequences; if based on supervision, it finds a breach of any obligations listed under letter (a) above;

c decision-making on stopping production or other activity (falling within the scope of its supervision competence) if it poses a very serious threat to the public interest and the environment;

d maintenance of a central register of accidents (§ 40 et seq. CzWL);

e carrying out control monitoring of the quality and quantity of discharged wastewater into surface water and conducting control of compliance by polluters with the provision of charges for the discharge of wastewater into surface water.

**River basin managers** (*správci povodí*) are legal entities established on the basis of Act No. 305/2000 Coll. (Act on River Basins, *zákon č. 305/2000 Sb., o povodích*) in the form of a state company (*státní podnik*). There are in total five river basin managers for five major river basins in Czechia: *Povodí Vltavy*, *Povodí Labe*, *Povodí Moravy*, *Povodí Ohře*, and *Povodí Odry* (Strnad et al., 2015). River basin managers are responsible for the management of major watercourses (§ 47 sect. 4 CzWL), activities related to the monitoring and assessment of the status of surface water and groundwater in a given river basin (§ 21 CzWL), and other activities listed in the CzWL such as the following:

1 Providing technical, economic and other available data for the activities of the water authorities and for the activities of the Czech Environmental Inspectorate (§ 54 sect. 3 CzWL)

2 Issuing opinion (*stanovisko*) to the requests for

   a water permit based on § 8 sect. lit. (a) to (c) CzWL, including the assessment of draft emission limit values for compliance with § 38 sect. 9 CzWL;

   b permit for an activity listed in § 14 sect. 1 lit. a) to (c) CzWL;

   c waterworks project permit (§ 55a CzWL);

   d binding opinion pursuant to § 17 CzWL or binding opinion pursuant to § 104 sect. 3 CzWL; and

   e exemption pursuant to § 23a sect. 8 CzWL, including an assessment of the possibility of deterioration in the status or ecological potential of the water body concerned by the project or the impossibility of achieving good status or good ecological potential of the water body concerned by the project (§ 23a sect. 7 CzWL).

3 Taking part in administrative proceedings concerning a major watercourse as a party to such proceedings (§ 115 sect. 5 CzWL)

4 Within the scope of their activities, submitting suggestions, proposals, and statements to water authorities and other administrative authorities in terms of the interests given by the relevant water plans and in terms of other interests pursued by law (§ 54 sect. 4 CzWL)
5 Cooperating with managers of small watercourses and authorised representatives of expert entities in solving problems concerning the whole river basin (§ 54 sect. 6 CzWL)

A **watercourse manager** (*správce vodního toku*) manages small watercourses (*drobné vody toky*), cares for watercourse channels, removes or transfers trees and bushes in adjacent land in cooperation with the landowners, and provides instruction on water use as part of the rules of handling a system of water reservoirs along a watercourse in an emergency (§ 45 CzWL).

## Surface water quality monitoring and assessment

Articles 4–8 of the WFD are implemented in § 21 and 22 CzWL, which regulate the monitoring and assessment of surface water (as well as groundwater) conditions (Horáček et al., 2015). There are also two implementing ministerial decrees which specify details of the monitoring and assessment of surface water condition, namely Decree No. 49/2011 Coll., on the Delimitation of Surface Water Bodies (*vyhláška č. 49/2011 Sb., o vymezení útvarů povrchových vod*), and Decree No. 98/2011 Coll., on the Method of Assessing the Status of Surface Water Bodies, the Method of Assessing the Ecological Potential of Heavily Influenced and Artificial Surface Water Bodies and the Details of Programmes for the Detection and Assessment of the Status of Surface Waters (*vyhláška č. 98/2011 Sb.*).

The monitoring and assessment of surface water condition are organised within the river basin and are carried out by river basin managers and expert bodies established by the competent ministries – namely, by the Czech Hydrometeorological Institute (*Český hydrometeorologický ústav*) and the T. G. Masaryk Research Institute of Water Management (*Výzkumný ústav vodohospodářský T. G. Masaryka, v.v.i.*). River basin managers and the aforementioned expert bodies also operate the relevant public administration information systems (§ 21 sect. 4 CzWL).

The monitoring and assessment of water condition pursuant to § 21 CzWL aim to provide the information basis for

- the exercise of public administration based on the CzWL (e.g., for compiling water balance information based on § 22 CzWL, for entering relevant information into public administration information systems maintained based on § 22 sect. 3–5 CzWL and for activities of river basin managers listed in § 54 sect. 4 CzWL);
- the water planning (§ 23–26 CzWL); and

- informing of public and of the competent authorities via *inter alia* obligation of the Ministry of Agriculture every three years to prepare, in cooperation with the Ministry of Environment and regional offices, and submit to the Government of the Czech Republic a summary report on the implementation of the programmes of measures and on the status of water and water management in the river basins (§ 26 sect. 7 CzWL; Horáček et al., 2015).

Pursuant to § 21 sect. 2 CzWL, the activities of monitoring and assessment comprise, in particular,

a monitoring of the quantity and quality of surface and groundwater, including their impact on human activities, and the monitoring of the status of water bodies and the ecological potential of heavily influenced and artificial water bodies;
b maintenance of water balance (*vodní bilance*, § 22 CzWL);
c establishment, maintenance and updating of records concerning objects and activities listed in § 21 sect. 2 lit. c) no. 1–14 CzWL (e.g., abstractions of surface water, discharge of wastewater and mine waters and their accumulation in inland reservoirs, various types of protected areas). Data from the records referred to in lit. (c) shall be published as open data.

As stipulated in § 21 sect. 3 CzWL, the guidelines for water bodies, definitions, and values of indicators and a system of monitoring of the status of water bodies, the mode for establishing the programme for monitoring and assessment of the water condition in individual parts of drainage basins, their contents and method for processing, methods applied, and the frequency of monitoring and other requirements for their application are laid down in Decree No. 98/2011, which was already mentioned earlier.

The Decree No. 98/2011 Coll. also specifies monitoring type and frequency. According to § 12 of Decree, the following programmes for identifying and assessing surface water condition are elaborated: (a) Framework Monitoring Programme (*Rámcový program monitoringu*); (b) Surface Water Monitoring Programme (*Program monitoringu povrchových vod*), which includes the Programme for Situational Monitoring of Surface Waters (*Program situačního monitoring povrchových vod*) and Surface Water Operational Monitoring Programme (*Program provozního monitoring povrchových vod*); (c) Monitoring Programme for Surface Water Quantitative Characteristics (*Program monitoring kvantitativních charakteristik povrchových vod*); and (d) Survey Monitoring Programme (*Program průzkumného monitoringu*).

As was already indicated earlier, the monitoring and assessment of surface water conditions are carried out by river basin managers who take the results thereof into account, especially when they deliver their opinions (*stanoviska*, § 54 sect. 4 CzWL) to requests for water permits based on § 8 or § 14 CzWL or to requests for binding opinions based on § 17 or 104 sect. 3 CzWL.

In their opinions, the river basin managers assess the possibility of deterioration in the status or ecological potential of the concerned water body or the impossibility of achieving good status or good ecological potential of the concerned water body. If the river basin manager arrives at a conclusion that it can be assumed that the proposed project, its change, or an activity will lead either to deterioration of the status or ecological potential of a surface water body (or groundwater body) or the impossibility of achieving good status or good ecological potential of a surface water body (or groundwater body), it indicates in its opinion that the project, its change, or an activity either requires an exemption pursuant to § 23a sect. 8 CzWL or that the project, its change, or an activity may not be carried out (exemption is not possible). Further detail on how the river basin managers shall proceed when assessing projects, their changes or activities are set in the Methodological Instruction of the Ministry of Agriculture and the Ministry of Environment on the assessment of the possibility of the impact of the project on the status of the affected water bodies (primary assessment) from 2018 (Ministerstvo zemědělství & Ministerstvo životního prostředí, 2018), which follows the methodical instruction from 2016.

The water authority (regional offices) may only grant the exemption pursuant to § 23a sect. 7 CzWL based on the application for an exemption (§ 23a sect. 9 CzWL) and provided that the following conditions set in § 23a sect. 8 CzWL are met:

- All practicable steps are taken to limit adverse effects of a project, its change or an activity on the status of the water body.
- The reasons for changes or modifications of the water body's status due to carrying out a project, its change, or an activity are in the overriding public interest or where the benefits to the environment and society in achieving the objectives referred to in § 23a sect. 1 CzWL are outweighed by the benefits of the new changes to human health, the maintenance of public protection, or sustainable development.
- The beneficial objectives resulting from the changes or modifications to the water body cannot, for reasons of technical impracticability or disproportionate cost, be achieved by other means which would be significantly better in environmental terms.

The water authority, which granted an exemption under § 23a sect. 8 CzWL must state and explain the reasons for granting every individual exemption in the applicable river basin management plan pursuant to § 24 CzWL or in its next update.

River basin management plans, as well as flood management plans, are reviewed and updated every six years since the day of their approval. The review and update of a river basin management plan proceeds in three consecutive phases, which also involve opportunities for public participation:

1 The preparatory works, which must include the following:

   a A timetable and work programme for the preparation of the river basin management plan, which must be published and made available for comment by water users and the public at least three years before the start of the period covered by the particular river basin management plan.
   b An analysis of the general and water management characteristics of the river basin, an assessment of the impacts of human activities on the status of surface water and groundwater, an economic analysis of water use, and a preliminary summary of significant water management problems identified in the river basin based on these, including the identification of artificial water bodies, the identification of heavily affected water bodies and the justification for it, and proposals for specific water protection objectives, which shall be published and made available for comment by water users and the public at least two years before the start of the period covered by the river basin management plan.

2 The preparation of a draft river basin management plan, which shall be prepared in accordance with the results of the preparatory works and shall include programmes of measures to achieve the objectives referred to in § 24 sect. 4 CzWL, published and made available to water users and the public for comments at least one year before the beginning of the period to which the river basin management plan relates.
3 The preparation of a river basin management plan adapted based on the evaluation of consultations with water users and the public.

Further details of the elaboration process are laid down in the ministerial Decree No. 50/2023 Coll.

The national river basin management plans for the three main river basins in the Czech Republic (Elbe or in Czech *Labe*, *Vltava*, and Danube or in Czech *Dunaj*) are elaborated by the Ministry of Agriculture in cooperation with the Ministry of Environment, and competent river basin managers and regional offices (§ 24 sect. 10 CzWL). The national river basin management plans are then approved by the Government of the Czech Republic and issued in the form of a measure of general nature (*opatření obecné povahy*) (§ 25 sect. 4 CzWL). National river basin management plans are subject to strategic environmental assessment, which is regulated by the Act No. 100/2001 Coll. (*zákon č. 100/2001 Sb., o posuzování vlivů na životní prostředí*).

Within each of the three main river basins, there are several sub-basins for which sub-basin management plans are elaborated by the competent river basin managers who cooperate with competent regional offices and central water authorities (§ 24 sect. 13 CzWL). The river Elbe (*Labe*) has five sub-basins (1. *Horní a střední Labe*, 2. *Horní Vltava*, 3. *Berounka*, 4. *Dolní*

*Vltava*, 5. *Ohře, Dolní Labe a ostatní přítoky Labe*), Oder (*Odra*) has two sub-basins (1. *Horní Odra*, 2. *Lužická Nisa a ostatní přítoky Odry*), and Danube (*Dunaj*) has three sub-basins (1. *Morava a přítohy Váhu*, 2. *Dyje*, 3. *Ostatní přítoky Dunaje*). Sub-basin management plans are also elaborated in the aforementioned procedure and issued by the regional offices in the form of a measure of general nature.

To complement the information on the law with the information from practice, in Table 2.1, we provide results of the surface water status monitoring for the years 2016–2018 and 2019–2021 conducted by the T. G. Masaryk Research Institute of Water Management, Biological Centre of the Czech Academy of Sciences and the Czech Hydrometeorological Institute (Mičaník et al., 2020; Vyskoč et al., 2024). The objectives set in the WFD and implemented in the § 23a CzWL were not achievable in 2015 (Chyba, 2014) and were not achieved in the following years after the deadline (22.12.2015). As the results of the monitoring show, the status of surface water gets worse in some characteristics than in previous years.

As presented by Vyskoč et al. (2024), the failure to achieve good chemical status during 2019–2021 was mainly due to the presence of polycyclic aromatic hydrocarbons (especially fluoranthene and benzo(a)pyrene). In the 'biota' matrix, mercury and brominated diphenyl ether are long-term problematic substances. Compared to the previous status assessment for the period 2016–2018, the presence of benzo[ghi]perylene and benzo[k]fluoranthene has gotten worse. In contrast, there was an improvement in the mercury assessment in the 'water' matrix. Good ecological status or potential was not achieved in 92.6% of rivers. The failure to achieve good ecological status/potential was mainly due to the status of biological components and the presence of total phosphorus. When compared to the previous status assessment for the period 2016–2018, there were no significant differences in the general physico-chemical indicators for the 'river' category, while the 'lake' category showed a deterioration in the assessment of transparency and an improvement in the assessment of the water reaction indicator.

*Table 2.1.* Chemical and Ecological Status of Surface Waters in Czechia

| *GOOD CHEMICAL STATUS failed to be achieved on % of* | *2016–2018* | *2019–2021* |
|---|---|---|
| - rivers | 51.1 % | 61.1% |
| - lakes | 20.5 % | 6.8 % |
| *GOOD ECOLOGICAL STATUS/POTENTIAL failed to be achieved on % of* | | |
| - rivers | 94.6 % | 92.6 % |
| - lakes | 86.3 % | 87.7 % |

*Resource:* Vyskoč et al., 2024

**Legal regulations concerning wastewater management**

The main sources of threats to surface water quality in terms of the CzWL are wastewater (*odpadní vody*, § 38 CzWL), hazardous substances (*závadné látky*, § 39 CzWL), and accidents (*havárie*, § 40 CzWL; Strnad et al., 2015).

An assessment of surface water quality is based on the ecological and chemical status and is closely connected to the regulation of wastewater contained in § 38 CzWL, which defines wastewater and specifies the rules of its discharge to waters and to soil.

Pursuant to § 38 sect. 1 CzWL, the **wastewater** is defined as the water used in residential, industrial, agricultural, medical, and other buildings, installations, or means of transport if it has an altered quality (composition or temperature) after use, and mixtures with rainwater, as well as other water flowing from such buildings, installations, or means of transport if it can pose a threat to the quality of surface waters and groundwater. Wastewater includes leachate arising from the operation of landfills and tailings ponds or their aftercare, except for waters that are used for the organisation's own use and waters that flow into mine waters.

The CzWL also contains a negative definition of wastewater in § 38 sect. 4. Wastewater under the CzWL is not (a) water from drainage systems of drained agricultural land; (b) cooling water used on vessels and for water turbines, which only raise the temperature; (c) unused mineral water from a natural medicinal source or a source of natural mineral water; and (d) rainwater from roads, if pollution of such water by pollutants is dealt with by technical measures in accordance with the ministerial decree implementing the Act on Roads (Decree No. 104/1997 Coll., *vyhláška č. 104/1997 Sb., kterou se provádí zákon o pozemních komunikacích*).

An entity which introduces wastewater to surface water is obliged to ensure its treatment in accordance with the conditions specified in the water permit, including the application of the best available technologies with respect to wastewater discharge (§ 38 sect. 5 CzWL) and to measure the volume of the discharged water and the level of its pollution (§ 38 sect. 6 CzWL).

When permitting the discharge of wastewater into surface or groundwater, the water authority shall, based on § 38 sect. 9 CzWL, set maximum permissible values for the quantity and concentration of discharged pollution (emission limits) and the volume of discharged water. In authorising the discharge of wastewater into surface waters, it is bound by the indicators reflecting the status of the water in the watercourse; environmental quality standards, indicators, and values of permissible pollution of surface waters; indicators and maximum permissible values of indicators of pollution of wastewater (emission standards) laid down by government regulation (Regulation No. 401/2015 Coll., *nařízení vlády č. 401/2015 Sb.*); and the terms and conditions of the wastewater discharge permit, including the specifications of the best available technologies in the field of wastewater disposal and the conditions for their use as laid down by the regulation and the best available techniques in the field of wastewater disposal.

Based on § 38 sect. 10 CzWL, the water authority shall, when permitting the discharge of wastewater into surface water (or groundwater), (a) take into account the need to achieve or maintain good status of surface or groundwater and water-related ecosystems and (b) consider the possibilities of limiting pollution at source and of limiting emissions to the environment as a whole and the possibilities of reusing wastewater. If the objectives set out in the relevant river basin management plan or the water protection objectives or environmental quality standards set out in a directly applicable EU regulation so require, the water authority shall set stricter emission limits in the water permit than the emission limits set out under § 38 sect. 9 CzWL, or may set additional indicators and their emission limits (§ 38 sect. 11 CzWL). The emission limits set by the water authority shall not be stricter than the values achievable using the best available techniques in the field of wastewater disposal.

The rules for handling **hazardous substances** are set in § 39 CzWL and in the ministerial decree on accidents (*havarijní vyhláška*, Decree No. 450/2005 Coll.). Based on § 39 sect. 1 CzWL hazardous substances are substances which are not wastewater or mine water and which may endanger the quality of surface water (or groundwater). Within hazardous substances, the CzWL differentiates also 'dangerous' hazardous substances (*nebezpečné závadné látky*) and 'especially dangerous' hazardous substances (*zvlášť nebezpečné závadné látky*).

Dangerous hazardous substances, especially dangerous hazardous substances and other substances or groups of substances of similar concern, are listed in Annex 1 to the CzWL. A special category of especially dangerous hazardous substances are the so-called priority substances (*prioritní látky*) which pose a significant risk to the aquatic environment and associated ecosystems. These priority substances are listed in the governmental decree (Decree No. 401/2015 Coll.), including the category of priority dangerous substances, which are substances that pose a very high risk in or via the aquatic environment due to their persistence and bioaccumulation capacity. In order to reduce pollution of surface and groundwater by hazardous pollutants and particularly hazardous pollutants, the government may, based on the proposal of the Ministry of Environment, adopt a Programme for the Reduction of Pollution of Surface and Groundwater by Hazardous Pollutants and Particularly Hazardous Pollutants (*Program na snížení znečištění povrchových a podzemních vod nebezpečnými závadnými látkami a zvlášť nebezpečnými závadnými látkami*) (§ 39 sect. 3 CzWL).

In relation to hazardous substances, the CzWL lays down a general preventative obligation of anyone who handles hazardous substances to take reasonable precautions to prevent these substances from entering surface water (or groundwater) and posing a threat to the environment (§ 39 sect. 1 CzWL). This general obligation is more specified in § 39 sect. 2 CzWL for cases when the user of hazardous substances handles these substances on a larger scale or where their handling is associated with an increased risk to surface water (or groundwater). The term 'larger scale' (*ve větším rozsahu*), in respect to handling hazardous substances, as well as 'handling hazardous substances associated with an increased risk to water' is defined in the Decree No. 450/2005 Coll. (§ 2 lit. (b) and (c)). In cases specified in § 39 sect. 2 CzWL, the user of the hazardous substances is obliged to (a) prepare an emergency plan (*havarijní plán*), which must be approved by the relevant water authorities and, after the approval, it must submitted to the integrated environmental reporting system (*integrovaný system plnění ohlašovacích povinností, ISPOP*), and (b) keep records of the measures taken and retain these records for five years.

More specific obligations are set for anyone who handles especially dangerous hazardous substances, dangerous hazardous substances, or hazardous substances on a larger scale or in cases where handling of hazardous substances is associated with increased danger. In these cases, § 39 sect. 4 and 6 CzWL lay down another set of specific obligations to take appropriate measures to ensure that hazardous substances do not enter the surface (or groundwater or drains which do not form part of the technological equipment of a production facility), *inter alia* obligations to

- install devices which capture, store, and process hazardous substances in a way that prevents them from entering the soil, mixing with wastewater or rainwater (§ 39 sect. 4 lit. (a) CzWL);
- use only such a device or method for handling hazardous substances which are also appropriate from the water quality protection perspective (§ 39 sect. 4 lit. (b) CzWL);
- establish a proper system of control for detecting substance leakage from defective substances (§ 39 sect. 4 lit. (e) CzWL); and
- maintain records of especially dangerous hazardous substances which are processed or used, their quantity, the content of active ingredients, properties, especially with respect to surface waters and to provide such information to water authorities on request and to the Fire Service of the Czech Republic (§ 39 sect. 6 CzWL).

Anyone who discharges wastewater to surface waters is obliged to pay a fee based on the amount of wastewater discharged. This specific category of fee is regulated in § 89–89q CzWL. According to § 89c CzWL, anyone who discharges wastewater to surface water is obliged to pay the fee when the relevant pollution indices are exceeded:

a Limit levels of mass and load concentration laid down in Appendix 2 to the CzWL (§ 89c sect. 2 CzWL)
b The amount of wastewater discharged exceeds 100,000 $m^3$ during a calendar year (§ 89c sect. 1 CzWL)

The fee is calculated as the product of the wastewater volume and the rate of 0.1 CZK/1 $m^3$. The fee for wastewater discharge is equal to the sum of partial amounts calculated according to individual pollution indices as the multiplicity of the fee rate and the total amount of wastewater discharged during the calendar year. The fees are paid to the bank account of the competent customs office and are counted as income of the State Environmental Protection Fund of the Czech Republic (§ 89k CzWL).

The CzWL regulates the **accidents** on waters in § 40–42. According to § 40 sect. 1 CzWL, an accident (*havárie*) denotes a particularly serious deterioration (*mimořádně závažné zhoršení*) or a particularly serious threat (*mimořádně závažné ohrožení*) to the quality of surface water (or groundwater). The CzWL (§ 39 sect. 2) further provides that the following situations also constitute an accident within the meaning of the CzWL:

- a A serious deterioration or a serious threat to the quality of surface water (or groundwater) by petroleum substances, dangerous hazardous substances, or radioactive emitters and radioactive waste
- b A deterioration or a threat to the quality of surface water (or groundwater) in protected area of natural water accumulation (*chráněná oblast přirozené akumulace vod – CHOPAV*) or in the protection zone of water resources (*ochranné pásmo vodního zdroje*)

Technical failures and defects in equipment for the capture, storage, disposal, transport, or disposal of hazardous substances which precede and are causally related to the accident are also considered to be part of the accident (§ 39 sect. 3 CzWL).

Obligations in case of an accident are laid down in § 41 CzWL, which has been significantly amended by the aforementioned emergency amendment of the CzWL (Act No. 182/2024 Coll.) as a result *inter alia* of the accident on the river Bečva in 2020. Handling the accident on the river Bečva and investigating its causes showed that the Czech law does not clearly divide competences of authorities so that they can effectively coordinate their actions in order to identify the polluter as well as the causes of the accident (Hlušičková & Vícha, 2024).

Based on the emergency amendment which came into force on 1 August 2024, the procedure of handling an accident is as follows. Whoever causes an accident or discovers an accident is obliged to report it immediately to the Fire Rescue Service of the Czech Republic ('FRS CZ'). If the person who caused the accident is known, the CzWL in § 41 sect. 2 obliges this person to follow the emergency plan (*havarijní plán*) or instructions of the competent water authority, of the Czech Environmental Inspectorate or of the FRS CZ and take immediate measures to eliminate the causes and consequences of the accident. Further, the person who caused the accident is obliged to cooperate with all the authorities who participate in accident clean-up work (§ 41 sect. 8 CzWL).

After the accident is reported to the FRS CZ, the FRS CZ immediately informs the competent water authority, the river basin manager, the Czech

Environmental Inspectorate, and the Police of the Czech Republic about the accident (§ 41 sect. 3 CzWL). In case the accident concerns sources of drinking water, the competent water authority informs immediately the Regional Hygiene Station (*krajská hygienická stanice*) about the accident. In case the accident occurred in the protection zone of a natural medicinal or natural mineral water source or on surface waters used for bathing, the Czech Environmental Inspectorate is responsible for immediately informing the Ministry of Health about the accident.

The water authority in whose administrative district the accident originated or where the accident was discovered is responsible for managing work on eliminating the causes and consequences of the accident (§ 41 sect. 4 CzWL) and investigating the accident (§ 41 sect. 5 CzWL). In case of an accident which requires extraordinary expertise, the Czech Environmental Inspectorate may take over from the water authority the management of the accident clean-up work and the investigation of the causes of accidents (§ 41 sect. 7 CzWL).

For the purposes of the accident investigation, the competent river basin manager ensures that the samples of accident-affected surface water (or groundwater), including relevant samples of sediments and living organisms and discharged wastewater in the accident-affected area, are taken and submitted without delay to a laboratory with appropriate accreditation for an analysis. Other authorities (such as the Czech Environmental Inspectorate) may also secure samples (§ 41 sect. 5 CzWL) if they decide to do so.

The Czech Environmental Inspectorate, the river basin managers, the Police of the Czech Republic, the Military Police, and the water authorities whose administrative districts were affected by the accident shall cooperate in the management of the accident clean-up work and in the investigation of the causes of the accident if called upon to do so by the FRS CZ or the competent water authority (§ 41 sect. 6 CzWL).

The CzWL lays down in § 42 the rules for taking **remedial measures** *(nápravná opatření)* in order to eliminate the effects of illegal wastewater discharge, illegal handling of hazardous substances, or accidents. Pursuant to § 42 sect. 1 CzWL, the competent water authorities or the Czech Environmental Inspectorate may order the perpetrator to carry out a remedial measure at their expense consisting in the elimination of the consequences of the defective condition or, if required by the nature of the matter, in the provision of an alternative source of water (Humlíčková, 2014; Strnad et al., 2015; Horáček et al., 2015). If the person to whom remedial measures have been imposed fails to comply with them and there is a risk of default, the competent water authority or the Czech Environmental Inspectorate shall ensure that the remedial measures are carried out at the expense of the person responsible for taking care of these measures.

The obligations arising from the remedial measures imposed on the originator of the defective condition or the acquirer of the property pursuant to § 42 sect. 2 shall pass to their legal successors (§ 42 sect. 3 CzWL). If remedial measures cannot be imposed and if there is a risk of serious threat or pollution of surface water (or groundwater), the competent water authority shall ensure necessary remedial measures on its own initiative or on the initiative of the Czech Environmental Inspectorate. It may, for this purpose, impose the implementation of remedial measures on a person with sufficient expertise to implement the remedial measures (§ 42 sect. 4 CzWL).

For the purpose of payment for remedial measures, the region (*kraj*) shall establish a special account within its budget, replenished annually up to 10 million CZK. The municipality with extended competence, whose water authority was competent to impose the implementation of the remedial measures, shall request the competent region (*kraj*) to cover the payment for the remedial measures implemented. The competent region shall then, without undue delay, send the funds in the agreed amount to the municipality with extended competence, which shall send them to the person who carried out the remedial measures (§ 42 sect. 4 CzWL).

Remedial measures pursuant to § 42 CzWL as described earlier shall not be imposed if a decision on the imposition of such measures pursuant to the Act No. 167/2008 Coll., on Prevention and Remediation of Ecological Damage and on Amendments to Certain Acts (*zákon č. 167/2008 Sb., o předcházení ekologické újmě a o její nápravě a o změně některých zákonů*) has been issued to remedy ecological damage to surface water (or groundwater). The Act No. 167/2008 Coll. implements the Directive 2004/35/EC on environmental liability with regard to the prevention and remedying of environmental damage. The special account of the region, established pursuant to § 42 sect. 4 CzWL, may also be used to cover the costs to remedy ecological damage to surface water (or groundwater) pursuant to Act No. 167/2008 Coll. (§ 42 sect. 5 CzWL).

## Administrative and penal responsibility of entities using surface waters

Water authorities, building authorities, and the Czech Environmental Inspectorate conduct water law supervision (*vodoprávní dozor*, § 110 CzWL) regarding compliance with obligations arising from the law, issued administrative decisions and measures of a general nature. The Ministry of Agriculture and the Ministry of Environment exercise the supreme water law supervision (*vrchní vodoprávní dozor*), which, pursuant to § 111 CzWL, means that they control compliance with the CzWL and other legislation by the water authorities and the Czech Environmental Inspectorate.

The Czech Environmental Inspectorate, water authorities, and building authorities decide on measures to be taken in order to remedy any deficits found during the water law supervision (§ 110 sect. 1 CzWL) and impose penalties for violations of the obligations of natural and legal persons laid down in the CzWL. When conducting water law supervision, the aforementioned authorities follow procedures and obligations laid down in the CzWL and in the Act No. 255/2012 Coll., on Control (*zákon č. 255/2012 Sb., o kontrole (kontrolní řád)*). Additionally, if the CzWL or the Act on Control does not contain the relevant procedural norms, the Administrative Procedure Act (Act No. 500/2004 Coll., *zákon č. 500/2004 Sb., správní řád*) as a *lex generalis* shall apply (Horáček et al., 2015).

The aforementioned authorities also impose sanctions for water law violations. In Czechia, administrative liability for offences (*odpovědnost za přestupky*) plays an important role in environmental protection, and there are several reasons for that (Jelínková & Havelková, 2018). First, the principle of subsidiarity of criminal liability applies (§ 12 sect. 2 of the Act No. 40/2009 Coll., Criminal Code, *zákon č. 40/2009 Sb., trestní zákoník*), which means that the criminal liability may only be invoked in socially harmful cases in which the application of administrative liability for offences is not sufficient. Second, in Czechia, criminal liability for environmental crimes is still applied only in a limited number of cases (Jelínková, Havelková, 2018). In Czech administrative practice, administrative liability for environmental offences is preferred, and environmental laws allow administrative authorities to impose relatively high monetary sanctions.

When deciding on environmental offences and imposing sanctions, the Czech Environmental Inspectorate, the water authorities (municipal offices of the municipalities with extended competence), and the building authorities follow the CzWL, which provides a list of environmental offences and lays down relevant sanctions for each type of offence in § 116–125k CzWL. Procedural rules are laid down in the CzWL (§ 125l) and in Act No. 250/2016 Coll., on Liability for Offences and Proceedings Thereon (*zákon č. 250/2016 Sb., o přestupcích a řízení o nich*).

Based on who the perpetrator is, the CzWL lists 80 offences of natural persons (§ 116–125) and 89 offences of business persons (§ 125a–125k), which may be committed by natural or legal persons during or as a result of their business activity.

A catalogue of offences of natural persons includes a general list of 21 offences in § 116 CzWL and specific groups of offences regarding (a) violations of obligations in the use of surface waters for navigation (§ 117); (b) violations of other obligations in the water management (§ 118); (c) violations of obligations of owners and builders of waterworks (§ 119); (d) violations of obligations of owners of land, buildings, and facilities (§ 120), (e) violations of obligations of owners in flood protection (§ 121); (f) violations of obligations in the handling of hazardous substances (§ 122); and (g) violations of obligations concerning accidents (§ 123), violations of obligations of a natural person who is the river basin manager of a small watercourse (§ 124), and violations of the fee obligation of a natural person (§ 125). The financial sanctions which may be imposed on a natural person range from 5,000 CZK to 1 million CZK.

A catalogue of offences of business persons includes a general list of 23 offences in § 125a CzWL and the same specific groups of offences as for natural persons but with higher sanctions. The financial sanctions which may be imposed on a business person range from 5,000 CZK to 5 million CZK.

To illustrate the differences between sanctions for offences of natural persons and business persons, we chose an example of two relevant offences. If a natural person uses surface water without authorisation, e.g., discharges wastewater into surface water without a permit, the competent authority may impose a fine up to 100,000 CZK (§ 116 sect. 1 lit. (b) and sect. 2 lit. (c) CzWL). If a business person commits the same offence, the competent authority may impose a fine of up to 500,000 CZK (§ 125a sect. 1 lit. (b) and sect. 2 lit. (b) CzWL).

If a natural person in contravention with § 39 sect. 1 CzWL fails to take reasonable precautions when handling hazardous substances in order to ensure that they do not enter surface water or groundwater and thereby endanger the water environment, the competent authority may impose a fine up to 500,000 CZK (§ 116 sect. 1 lit. (l) and sect. 2 lit. (d) CzWL). In case a business person commits the same offence, the competent authority may impose a fine of up to 1 million CZK (§ 125a sect. 1 lit. (m) and sect. 2 lit. (e) CzWL). In determining the amount of the fine for an offence committed by discharging wastewater without a water permit or in contravention thereof, the authority imposing the fine shall take into account, in particular, the extent to which the conditions of the permit for the discharge of such water have been exceeded, the extent of the impact on the quality of surface water or groundwater and its local extent, the degree of protection of the area concerned and the cause of the unauthorised discharge (§ 125l sect. 3 CzWL).

Based on § 125l sect. 2 CzWL, the rate of a fine for an offence imposed under the CzWL shall be doubled but shall not exceed 20 million CZK if the same offence is committed repeatedly. An offence is committed repeatedly if one year has not elapsed since the date on which the decision to impose a fine for the same offence became final.

What concerns criminal liability regarding water protection? Environmental crimes are listed in Section VIII of the Criminal Code (§ 293–308), which implements the relevant EU law. In relation to surface water protection, the following crimes listed in the Criminal Code are relevant (Pekárek, 2015):

- intentional or negligent damage or endangerment of the environment (§ 293 and 294);
- damage of water resource (§ 294a);
- unauthorised discharge of polluting substances (§ 297);
- unauthorised management of waste, which causes damage or endangerment of *inter alia* surface water (§ 298);
- unauthorised handling of protected species of wild animals and plants (§ 299 and 300); and
- damage of surface waters which are protected by the Act on Nature and Landscape Protection (Act No. 114/1992 Coll., *zákon č. 114/1992 Sb., o ochraně přírody a krajiny*) (§ 301).

The main deficiencies of the Czech criminal justice system concerning environmental crime may be summarised as follows (Kotlán, 2024): lack of acceptance and awareness regarding the seriousness of environmental crime, lack of proper expert training of the criminal justice professionals, shortage of funds as well as specialised personnel (or even specialised units within the police force, public prosecution, and courts), high rate of environmental crime latency as well as low clearance rate of registered suspicions, disproportionately lenient punishment and ineffective criminal liability and punishment of corporations.

Corporate criminal liability was introduced into the Czech legal order based on *inter alia* implementation of the Directive 2008/99/EC of the European Parliament and of the Council of 19 November 2008 on the protection of the environment through criminal law. The Act No. 418/2011 Coll., on Criminal Liability of the Legal Persons and Proceedings Against Them (*zákon č. 418/2011 Sb., o trestní odpovědnosti právnických osob a řízení proti nim*) came into force on 1 January 2012, two years after the transposition deadline (26 December 2010). This new institute of corporate criminal liability was introduced with some level of distrust, and the lack of its acceptance still affects the effectiveness of the Czech criminal justice system when it comes to investigating and prosecuting environmental crimes conducted by corporations (Kotlán, 2024). The implementation of the new Directive (EU) 2024/1203 of the European Parliament and of the Council of 11 April 2024 on the protection of the environment through criminal law, replacing Directives 2008/99/EC and 2009/123/EC will definitely constitute a new (renewed) challenge for the Czech legislator, as well as for all law enforcement authorities.

## Summary

As an EU Member State, Czechia is responsible for implementing the EU law requirements for surface water quality, as well as for achieving objectives set in the WFD and related directives. For a country with far too many small municipalities (88.6% of Czech municipalities have less than 2,000 inhabitants), with a mixed system of public administration (state administration tasks are executed by the state organs, as well as by self-governing regions and municipalities as delegated state administration), with several ministries sharing competencies in the area of water protection and water management, the achievement of surface water protection objectives not only from WFD has proved to be difficult. The objective to achieve a good chemical status of surface waters was unachievable by the deadline set in the WFD (22 December 2015) mainly due to phosphorus contamination (Chyba, 2014). As the surface water status monitoring in the years 2019–2021 showed, Czechia failed to achieve a good chemical status on 61 % of its rivers. The objective of a good ecological status/potential failed to be achieved on nearly 93 % of the Czech rivers (Mičaník et al., 2020; Vyskoč et al., 2024). Even ten years after the deadline set by the WFD, the objectives are not met.

The Czech water law represented in the first place by Act No. 254/2001 Coll. (and complemented by several other laws, governmental regulations, and ministerial decrees) specifies in detail the environmental goals in line with the WFD. The Czech water law contains a complex mix of instruments for achieving all the goals and, with them, effective protection of surface water, including a water management register, water planning, monitoring and assessment of the water quality and quantity, permits for surface water management and a public administration information system, and provides a system of rules for handling accidents and imposing remedial measures. Moreover, the Czech water law pays particular attention to the responsibilities of entities which use water for industrial purposes to adapt production processes with a view to the economical use of water resources with the use of the best available technology. Particular emphasis is also placed on water supply and proper treatment of wastewater produced during a construction process, and as a rule, every new building constructed must have a proper water supply and wastewater treatment.

An assessment of surface water quality is based on the ecological and chemical status of the waters. The Czech legal system provides for a system of surface water monitoring and assessment within which the competent river basin managers, based on the results of monitoring and assessment, provide their opinions for the purposes of each water proceedings within which the water authorities decide on requests for water permits.

The Czech water law establishes a possibility to change water use rules laid down in a permit (§ 12 sect. 3 CzWL) should the emission norms for pollutants introduced to waters change, with a view to meeting the conditions of reduction programmes for surface water pollution (*programy snížení znečištění povrchových vod*), including dangerous hazardous substances (*nebezpečné závadné látky*). Interestingly, the legislation allows for the discharge of contaminated groundwater to surface water in order to reduce the level of its contamination. For such an activity, a water permit pursuant to § 8 sect. 1 lit. (e) CzWL is necessary, not a water permit for an ordinary discharging of wastewater pursuant to § 8 sect. 1 lit. (c) CzWL. Moreover, the definition of wastewater does not include cooling-down water from turbines, which is only characterised by an elevated temperature, and mineral waters from natural sources.

The CzWL also includes rules for handling hazardous substances, including in emergencies, which enter surface waters and pose a threat to the aquatic environment (§ 39 of the CzWL). The course of handling the aforementioned accident on the river Bečva and the subsequent investigation of the causes of this accident fully revealed the shortcomings of the Czech legislation, the fragmentation of which did not allow a coordinated procedure leading to an effective investigation and prosecution (Kotlán, 2024). Under immense pressure from media and public attention, the Ministry of Environment prepared the so-called emergency amendment of the CzWL, which was finally passed by the Czech Parliament in June 2024 (Act No.

182/2024 Coll.). The emergency amendment introduced new instruments (such as a register of discharges from pollution sources into surface waters) and clarified competencies of the authorities under § 41 CzWL. Before the emergency amendment, the CzWL allowed for overlapping competencies during the handling of an accident, which in the case of the Bečva accident resulted in delayed reaction and, as a result thereof, ineffective investigation. Despite the clarification brought by the emergency amendment, the Czech water law remains fragmented and scattered in too many legislative acts, allowing for gaps and unnecessary overlaps, starting with four ministries on the central level being competent in the area of water law.

Besides ministries, the surface water management and protection tasks are carried out by water authorities *(vodoprávní* úřady) on regional and local levels in close cooperation with the Czech Environmental Inspectorate. Due to the mixed system of public administration, the water authority is exercised by the territorial self-governing bodies as a task of delegated state administration. The water permitting procedure (*vodoprávní řízení*) is mostly carried out by the municipal offices of 205 municipalities with extended competence (*obecní úřady obcí s rozšířenou působností*), these water authorities issue other consents and declarations concerning construction processes, take actions in emergencies, etc. Apart from already mentioned authorities, there are appointed river basin managers (*správci povodí*) established by the law in form of a state company (*státní podnik*), who are responsible for managing drainage basins and important watercourses, and watercourse managers (*správci vodních toků*), who manage small watercourses, which streamlines the management of such areas by strict control of the water quality and the type of their use, especially for industrial purposes.

While enforcement of administrative liability for environmental offences in Czechia is generally considered to be sufficient, criminal liability and prosecution of environmental crimes, especially those conducted by persons engaged in business (legal as well as natural persons), remains insufficient (Kotlán, 2024). In Czechia, the criminal punishment of especially economic environmental crime is ineffective and not fulfilling the potential and role which the criminal liability should play in a democratic legal order where the state holds constitutional obligation to protect the environment (Art. 7 of the Czech Constitution) and protect fundamental rights of everyone to health and favourable living environment (Art. 31 and 35 of the Czech Charter of Fundamental Rights and Freedoms).

## Notes

1 There are four military districts (*vojenské újezdy*) in Czechia: Boletice, Březina, Hradiště, and Libavá.

2 There are 6,258 municipalities in Czechia. Only 5 municipalities have more than 100,000 inhabitants, 122 municipalities have between 10,000 and 100,000 inhabitants. There are 584 municipalities with a number of inhabitants between 2,000 and 9,999. There are 778 municipalities with a number of inhabitants

between 1,000 and 1,999. There 1,375 municipalities with a number of inhabitants between 999 and 500 and 3,394 municipalities with less than 500 inhabitants. In total, 88.6 % of the Czech municipalities have less than 2,000 inhabitants (Gremlica, 2023).

## References

Česká společnost pro právo životního prostředí (2021). Vyjádření České společnosti pro právo životního prostředí k aplikaci zákona č. 167/2008 Sb., o předcházení ekologické újmě a o její nápravě, na případ znečištění řeky Bečvy kyanidy. *České parvo životního prostředí* 60 (2):69–79.

Chyba, J. (2014). Je dobrý stav povrchových vod reálně dosažitelný? In *Voda v právních vztazích, Spisy Právnické Fakulty Masarykovy Univerzity, řada teoretická, č. 481*, edited by I. Průchová and J. Hanák, et al., 133–139. Masarykova univerzita.

Gremlica, T. (2023). Fundamental revision of the Urban Waste Water Treatment Directive provokes conflicting reactions from European Union member states. *Water Management Technical and Economical Information Journal* 65 (4):52–53.

Hlušičková, D., and O. Vícha (2024). Přeshraniční znečištění v případu úniku kyanidu z těžebního provozu v Baia Mare: Aspekty unijního a českého práva (Vol. II). *Acta Universitatis Carolinae – Iuridica* 70 (1):93–109.

Horáček, J. et al. (2015). *Vodní zákon s aktualizovaným podrobným komentářem po roce účinnosti nového občanského zákoníku k 15. 3. 2015*. Sondy.

Humlíčková, P. (2014). Odpovědnost za ztráty na vodách. In *Voda v právních vztazích, Spisy Právnické Fakulty Masarykovy Univerzity, řada teoretická, č. 481*, edited by I. Průchová and J. Hanák, et al., 190–200. Masarykova univerzita.

Huneš, K., and V. Tomoszková, et al. (2022). *Veřejné užívání přírodních statků*. C. H. Beck.

Jelínková, J., and S. Havelková (2018). *Přestupky v praxi orgánů ochrany životního prostředí*. C. H. Beck.

Kopecký, M. (2023). *Správní právo Obecná část*. C. H. Beck.

Kotlán, P. (2024). Problems of sanctioning economic environmental crime in the Czech Republic. *Journal of Economic Criminology* 3:1–6. https://doi.org/10.1016/j.jeconc.2024.100050

Mičaník, T., et al. (2020). Surface Water State Assessment for Third Cycle River Basin Management Plan of the Czech Republic. *Water Management Technical and Economical Information Journal* 62 (6):4–18. https://doi.org/10.46555/VTEI.2020.09.002

Ministerstvo zemědělství & Ministerstvo životního prostředí (2018). Metodický pokyn sekce vodního hospodářství Ministerstva zemědělství a sekce technické ochrany životního prostředí Ministerstva životního prostředí k posouzení možnosti vlivu záměru na stav dotčených vodních útvarů (primární posouzení) [§ 23a zákona č. 254/2001 Sb., o vodách a o změně některých zákonů (vodní zákon), ve znění pozdějších předpisů]. https://www.mzp.cz/C1257458002F0DC7/cz/vodni_utvary_metoika/$FILE/OOV-MP_vodni_utvary-20180301.pdf

Müllerová, H. (2015). *Legal regulation of natural resources in the Czech Republic: an overview. Working Paper Nr. 9/2015*, 1–30. Academy of Sciences of the Czech Republic. http://doi.org/10.2139/ssrn.2514190

Pekárek, M. (2015). I.2 Ochrana vod a vodních zdrojů. In *Právo životního prostředí: zvláštní část*, edited by I. Jančářová, et al., 88–162. Masarykova univerzita.

Průcha, P. (2008). Samostatná a přenesená působnost místní (a regionální) samosprávy (smíšený model územní veřejné správy). In *Právní regulace místní (a regionální) samosprávy: sborník z 4. letní mezinárodní conference/workshopu. Kroměříž, Právnická fakulta Masarykovy university a Veřejný ochránce práv, 19. – 20. června 2008*, 203–220. Masarykova universita.

Sládeček, V. (2019). *Obecné správní právo*. Wolters Kluwer.
Sobotka, M. (2010). Kapitola VII. Ochrana složek životního prostředí. Oddíl 2. Ochrana vody. In *Právo životního prostředí*, edited by M. Damohorský, et al., 275–294. C. H. Beck.
Staša, J. (2012). Kapitola XIV. Veřejné užívání. In *Správní právo. Obecná část*, edited by D. Hendrych, et al., 307–319. C. H. Beck.
Strnad, Z., et al. (2015). *Vodní právo*. Jihočeská univerzita v Českých Budějovicích, Fakulta rybářství a ochrany vod.
Vomáčka, V., and J. Tkáčiková (2022). Agricultural Issues and the Czech Constitution. *Journal of Agricultural and Environmental Law* 32 (1):151–171. https://doi.org/10.21029/JAEL.2022.32.157
Vomáčka, V., M. Tomoszek, and V. Tomoszková (2021). Komentář k čl. 35. In *Listina základních práv a svobod. Komentář*, edited by F. Husseini, et al., 974–1031. C. H. Beck.
Vyskoč, P., et al. (2024). Assessment of the status of surface water bodies in the Czech Republic for the period 2019–2021. *Water Management Technical and Economical Information Journal* 66 (3):10–18. https://doi.org/10.46555/10.46555/vtei.2024.03.002
Žídek, D. (2021). Environmental protection in the Constitution of the Czech Republic. *Journal of Agricultural and Environmental Law* 16 (31):145–160. https://doi.org/10.21029/JAEL.2021.31.145

# 3 Legal protection of surface waters in France

*Michał Mariański and Richard Bartes*

## Introduction

The analysis presented in this part of the study deals with legal regulations applicable to surface water management and protection in France. This part of the study presents a review of legal acts from the field of surface water management and protection in France, as well as legal acts at the foundation of the water protection system, as well as the bodies/institutions established on the basis of such regulations. Water resources are often compared to oil, the shortage of which would lead to an imminent, permanent, local, and global 'water crisis'. More than a 'water crisis' linked to a natural shortage, water is an important element of local and worldwide geopolitics as it is governed by deficient water policies and by the difficulty of guaranteeing water security. In this context, we can observe several common national major challenges for water management in the 21st century: preserving the ecosystem, providing drinking water to all, and providing sufficient water for agriculture (Blanchon, 2024).

However, it must be pointed out that the French legal system is quite specific in its organisational, administrative, and legal aspects. Thus, for example, French local governments are usually referred to as the territorial administration (*administration teritoriale*) or decentralised administration (*administration décentralisée*) and have a significant impact not only on the development of elements but also on the state as a whole. Their importance increased significantly after the Act of Liberties of Communes, Departments and Regions of 1982 was passed (the Act of 2 mars 1982 No. 82-213 – *Loi relative aux droits et libertes des communes, des departements et des regions*), which introduced a division into three basic territorial units: communes, departments (*départements*) and regions (*régions*). The specificity of the French model shows in other defined special status units (*collectivité à statut particulier*) and overseas collectivities (*collectivité d'outre-mer*; Mariański, 2021).

It is noteworthy that the French constitution has been using the uniform term of 'territorial collectivity' (*collectivité territoriale*) since its revision of 28 March 2003, which is consistent with the regulations laid down in the General Code of Territorial Collectivities (*Code général des collectivités territoriales*).

DOI: 10.4324/9781003586913-4

The Water Framework Directive in France is referred to as the DCE (fr. *Directive-Cadre sur l'Eau*) and was implemented into the French law by the Act No. 2004-338 of 21 April 2004, transposing Directive 2000/60/EC of the European Parliament and of the Council of 23 October 2000 establishing a framework for community action in the field of water policy, which was published in *Journal of Laws* No 95 of 22 April 2004, page 7327 (*fr. loi n°2004-338 du 21 avril 2004 portant transposition de la directive 2000/60/ CE du Parlement européen et du Conseil du 23 octobre 2000 établissant un cadre pour une politique communautaire dans le domaine de l'eau*). Nevertheless, the essential legal act in this study – along with the aforementioned code – will be the Code of Environment (*Fr. Code de l'environnement*), and, in particular, its Part II, entitled 'Waters and Aquatic and Marine Environment', which corresponds to Articles L210-1 to L219-18 of this code.

Before providing a detailed analysis of this legal act, it is worth pointing out that the French domestic law passed in the form of codes is of particular importance for the French legal system. In the French legal system, codes regulate practically every area of the law; they emphasise their importance, but in practice, they address specific issues in one legal act. Moreover, unlike codes passed in other legal systems, French laws have a distinctive structure, which takes into account not only the legal norms based on acts and regulations but also elements of case law and doctrine in a specific branch of the law. Thus, the structure of most codes allows for creating a uniform document for a specific area of law, which basically consists of legislative (L) and regulative (R) parts (Klimaszewska et al., 2017).

It is worth noting, however, that the role of the highest administrative court (*Conseil d'Etat*) remains very important because this institution can clarify statutory provisions by way of decrees having legal force. It is worth noting here that in the French legal order, there are three categories of dispositive decrees (*décrets réglementaires*): ordinary decrees (*les décrets simples*), decrees of the highest administrative court (*les décrets en Conseil d'État*) and decrees of the Council of Ministers (*décrets en conseil des ministres*; Fabre-Magnan, 2014). In this aspect, it is important to underline the quite specific role played by the highest administrative court in the French judiciary. In other legal orders, such as in Poland, the Supreme Administrative Court does not have such extensive competences. The doctrine also points to the fact that, conceptually, the administrative judiciary system in France has a structure resembling a planetary system, where the key element and the centre around which the other elements are organised is the highest administrative court – Conseil d'Etat (Chapus, 1999).

## Legal basis of surface water management and protection

Historically, water management and protection are part of environmental law (*Fr. droit de l'environnement*) related to the ecological transition. The ecological transition is the evolutionary process where we can observe the contradiction between human activity and the preservation of natural elements being also characterised by the complexity of the economic and social relationships. In order to create an appropriate national system of water management and protection, a combined perspective of law and economics is required, as well as an institutional framework created by the state (Martin & Gaffard, 2023). It is considered that a comprehensive reference instrument was given to French environmental law when it was fully codified in 2000 and took into consideration all the necessary actors related to the environment, like associations, local authorities and their groups, legal professions, administrations, industrialists, farmers, students, and, finally, ordinary citizens (Prieur, 2023). In France, the term 'public water policy' (*fr. La politique publique de l'eau*) is very often used – that is, related to the fact that the question of water, its availability for the present and in the future, will be increasingly marked by the impacts of climate change and needs public protection. As a consequence, many cities in France have chosen to introduce or to return to public water management that is based on the idea that water is a common good and thus needs a water governance system (Defeuilley, 2024). The legal basis that will be described in this part of the study was created in order to ensure water – in all its forms and components – as a model of common good supported by legal regulations likely to improve access, sharing, governance, and protection of this vital resource.

The legal basis related to water management and protection may be divided into two parts. The first part is related to classical main legal acts (laws), and the second group consists of the subsidiary of 18 decrees that are instruments detailing the provisions of the main regulations regrouped in the first part.

As for the first part of the legal basis in the field of water protection, we can enumerate several acts of law. Chronologically, the first was the oldest act, No. 64-1245 of 16 December 1964 on water regime and distribution and on combating its pollution. Second, Act No. 92-3 of 3 January 1992 on water (Articles L212-3 to L212-7 of the Code of Environment), and third, Act No. 2000-108 of 10 February 2000 on modernisation and development of public electric service. Fourth, Act No. 2004-338 of 21 April 2004 – the most important one concerning these issues in the French legal system, introducing the most current provisions to the Code of Environment (*Fr. Code de l'environnement*), transposing Directive 2000/60/EC of 23 October 2000 establishing a framework for community action in the field of water policy. Fifth, Act No. 2006-1772 of 30 December 2006 on waters and aquatic environment. Sixth, programme Act No. 2009-967 of 3 August 2009 on holding Grenelle – the environmental forum. Seventh, Act No. 2010-788 of 12 July 2010 on the state's commitment to protecting the environment and another Act No. 2014-58 of 27 January 2014 on modernisation of territorial public actions and affirmation of a metropolis – so-called MAPTAM. Ninth, Act No. 2015-991 of 7 August 2015 on the new territorial organisation of the Republic. The last, chronologically tenth – i.e., the latest legal act – is Act No. 2016-1087 of 8 August 2016 on restoring biodiversity, nature, and landscapes.

As for the first second part of the legal basis in the field of water protection, we can enumerate 18 subsidiary decrees regulating this subject matter, of which the first was issued in 1992 and the latest in 2018. Like with the acts of law, the oldest one is the Decree No. 92-1042 of 24 September 1992, executing Article 5 of Act No. 92-3 of 3 January 1992 on water and concerning plans of development and management of waters, and amended with the Decree No. 2005-1329 of 21 October 2005. The next legal act of this type is Decree No. 2005-475 of 16 May 2005 on plans of development and management of waters, after which Decree No. 2005-613 of 27 May 2005 was added to the assessment of the environmental impact of some plans and programmes and Decree No. 2005-636 of 30 May 2005 on organisation of water management administration and the mission of the prefect coordinator for the drainage basin. Subsequent decrees modified selected articles in the Code of Environment, i.e., No. 2005-1329 of 21 October 2005, based on Articles L.212-3 to L.212-7 of the Code of Environment, Decree No. 2006-880 of 17 July 2006 on procedures of granting permits and declarations provided for in the Code of Environment, and No. 2007-135 of 30 January 2007 specifying the criteria for defining and delimiting water bodies and wetlands mentioned in art. L. 211-1 of the Code of Environment. The eighth decree was No. 2007-882 of 14 May 2007 on some areas subject to environmental constraints and modifying the agricultural code. Another four decrees were passed in 2007: No. 2007-1213 of 10 August 2007 on plans of development and management of waters, and modifying the Code of Environment; No. 2007-1557 of 2 November 2007 on the basic nuclear facilities and on control of nuclear safety of transport of radioactive substances; No. 2007-1760 of 14 December 2007 establishing regulations concerning a system of permits and declarations on management and protection of waters and aquatic environment, obligations applicable to some works located on watercourses, maintaining and restoration of aquatic environments; and an amendment of the Code of Environment and the Decree No. 2007-1872 of 26 December 2007 on intention to use artificial watercourse flow for specific purposes. Subsequently, the next two decrees were added in 2011: No. 2011-2018 of 29 December 2011 on reform of opinion polls on actions with a potential environmental impact and No. 2011-2021 of 29 December 2011 on a list of projects, plans, and programmes communicated to the public via electronic means. Decree No. 15 on the issue in question was issued in 2012, i.e., No. 2012-616 of 2 May 2012 on the assessment of some plans and documents with environmental impact, and three years later, Decree No. 2015-1038 of 20 August 2015 on regional public water bodies and public water management facilities. Subsequently, in 2017, a rather important decree, No. 2016-1071 of 3 August 2016, was issued on local plans, sustainable development, and territorial equality, and the 18th – the latest – legal act of this type was Decree No. 2018-847 of 4 October 2018 on guidelines concerning water management plans.

As was mentioned in the introduction to this subchapter, the legal acts mentioned earlier, including laws and decrees, are basically reflected in the Code of Environment (*Fr. Code de l'environnement*) and, as an auxiliary measure, in the General Code of Territorial Collectivities (*Code général des collectivités territoriales*).

Historically, the code Code of Environment (*Fr. Code de l'environnement*) brought together in France all the laws and directives relating to environmental law. Seven books defined all the measures applicable in mainland France and in the overseas departments and territories for the preservation of environments and biodiversity. The code was created in September 2000 and underwent two modifications in 2005 and 2007. It has direct links to other legal acts, like to urban code, in particular, on the use of land and the preservation of coastal and mountain areas, and to the Commercial Code, on the role of economic activities and the risks they present for the environment.

The Environmental Code defines four basic principles for reasoned and sustainable management of natural environments and, in particular, in the case of water management and protection. First, the precautionary principle states that we must not wait for a natural disaster or major pollution to control environmental risks. Effective measures must be taken for the sake of prevention and not in reaction to a crisis: the absence of certainties does not justify, therefore, the passivity of public policies. Second, the polluter-pays principle is a tool for combating pollution and the various pressures that humans impose on the environment. It requires those responsible for major pollution to pay for and repair the damage caused themselves. Third, the principle of preventive and corrective action is a legal system whose role is to find the best solutions to repair damage to the environment with a respectable economic cost. Fourth, the principle of participation invites all citizens to contribute to the preservation of natural environments and French biodiversity. According to him, everyone must have access to information and data relating to chemical, nuclear, or biological risks and to all dangerous activities in general (Roche, 2022).

The key regulations within the Code of Environment are those contained in Title I of Book II, 'Waters and Aquatic and Marine Environment', which corresponds to Articles L210-1 to L219-18 of the aforementioned legal act. The title consists of nine consecutive chapters, which regulate, in sequence: general system and resource management in Chapter I, i.e., in Articles L211-1 to L211-14 (*fr. Chapitre Ier: Régime général et gestion de la resource*); planning in Chapter II, i.e., in Articles L212-1 to L212-11 (*Fr. Chapitre II: Planification*); administrative and financial structures in Chapter III, i.e., in Articles L213-1 to L213-22 (*fr. Chapitre III: Structures administratives et financières*); activities, equipment and use in Chapter IV, i.e., in Articles L214-1 to L214-19 (*fr. Chapitre IV: Activités, installations et usage*); specific provisions concerning non-state watercourses in Chapter V and Articles L215-1 to L215-18 (*fr. Chapitre V: Dispositions propres aux cours d'eau non domaniaux*); provisions concerning control and penalties in Chapter VI, i.e., in Articles L216-1 to L216-13 (*fr. Chapitre VI: Dispositions relatives aux contrôles et sanctions*); issues related to national defence in Chapter VII and Articles L217-1 to L217-3 (*fr. Chapitre VII: Défense nationale*); specific regulations concerning marine waters and routes open for marine navigation in Chapter VIII, i.e., Articles L218-1 to L218-86 (*fr. Chapitre VIII: Dispositions spéciales aux eaux marines et aux voies ouvertes à la navigation maritime*); and policy concerning marine environments in Chapter IX, i.e., in Articles L219-1 A to L219-18 (*fr. Chapitre IX: Politiques pour les milieux marins*).

The General Code of Territorial Collectivities (*Code général des collectivités territoriales*), which is supplementary to these regulations and affects the way the competences in this regard are executed by decentralised units, contains only three groups of articles, i.e., Articles L4424-36 concerning special dispositions for Corsica, Articles L5421-1 à L5421-6 concerning interdepartmental institutions and organisations (*Fr. institutions et organismes interdépartementaux*), and Articles L5711-1 à L5721-7 concerning so-called mixed syndicates.

### Water authority – structure, responsibilities, and competences

The public and social interest was often the basis used by French lawmakers to justify the introduction of new legal authorities in the field of water protection. The development of such authorities, from public law entities through local authorities, shows that along with the evolution of the concept of the scope of water protection, there was also an evolution of the institutional aspect in this field. Originally, French territorial units for a long time were based on two typical French legal concepts of public undertakings that could support the realisation of an environmental policy – public establishments (fr. *établissement public*) and public enterprises (fr. *entreprise public*). Also, the notion of public service and public interest appeared in French law through the case law of the Highest Supreme Administrative Court – *Conseil d'Etat*. It is particularly noteworthy that the meaning and extent of the concept of public service was explained by the French judiciary in the Blanco judgement dated 8 February 1873 and in the Pelltier judgement dated 30 July 1873 in which the responsibilities of public authorities were defined.

The development of the institutional aspect of water protection was due to the need to perform many tasks in the general social interest. Therefore, it was decided to create a special category of institutions in this field whose structure was based on the provisions of French administrative and commercial law. It was noted that modifying the structure of a public company may help French territorial units create quasi-public entities which more easily and flexibly could perform public tasks.

Finally, that is in Chapter III of the French Environmental Code where we can find the provisions related to the institutional aspect of water protection (*Chapitre III: Structures administratives et financières*). This chapter includes Articles L213-1 to L213-22, where the Title I of Book II (Waters and Aquatic and Marine Environment) is of key importance to the institutional model of surface water management and protection in France.

This chapter consists of several sections that were modified by the laws and decrees mentioned in the previous section as part of the evolution of the French model in this regard. In order to show this evolution, institutions which have been removed from the French legal system will also be mentioned, which demonstrates a tendency to simplify its structure. Therefore, there are no longer such institutions as the National Bureau for Waters and Aquatic Environments (*Fr. Office national de l'eau et des milieux aquatiques*), which was abolished by the 2016 law (LOI n° 2016-1087 du 8 août 2016 pour la reconquête de la biodiversité, de la nature et des paysages (1)). The original regulations on so-called water agencies (*Fr. Agences de l'eau*) and the National Water Fund (*Fr. Fonds national de l'eau*) were also repealed. Local water communities (*Fr. Communautés locales de l'eau*) and public institutions with a vocation for project management, formerly in Sections 6 and 7 of this section (*Fr. Etablissements publics à vocation de maîtrise d'ouvrage*), and project management organisations (*Fr. Organismes à vocation de maîtrise d'ouvrage*) were also liquidated. Local water communities (*Fr. Organismes à vocation de maîtrise d'ouvrage*).

Thus, the first institution is the National Committee of Water, regulated in Articles L213-1 of the Code of Environment (Fr. *Comité national de l'eau*). In light of the provision referred to earlier, the National Water Committee is responsible for issuing opinions regarding the geographic borders of basins (waterbodies, drainage basins) and groups of basins (waterbodies, drainage basins) and any problems common to two or more areas of this kind; issuing opinions on all the domestic water management and distribution projects, as well as important regional events; issuing opinions on regulations concerning the protection of fish and crustacean populations; and issuing at the request of the advisory committee appointed as part of it opinions on the price of water charged from users and the quality of water supply and sewage services. Moreover, characteristically for French law, the detailed composition, organisation, and operation of the committee are defined by the relevant decree.

Another institution is the prefect coordinator for water bodies (*Fr. Préfet coordonnateur de bassin*), whose competences are regulated in Articles L.213-7 of the Code of Environment. According to this article, the prefect of the region, in which the committee of the drainage basin has its seat, manages and coordinates the state policy with respect to the water policy and management in order to achieve the coherence and uniformity of the decentralised state actions in this area within the limits of regions and departments. Moreover, the prefect follows and coordinates the state policy regarding the assessment of and management of flood risk, referred to in Chapter VI of Title VI of Book V. The decrees provided for in Article L. 211-2 specify the conditions of interventions of the river basin prefect coordinator, in particular with respect to crisis situation management, as well as all kinds of measures necessary to perform the mission. It should be emphasised that the institution of the prefect in French law is close to that of the voivode in Polish law.

The third institution includes the committees for river basins and water agencies (*Fr. Comités de bassin et agences de l'eau*), which are regulated in Articles L213-8 to L213-11-17 of the Code of Environment. The general provisions in this section are laid down in Articles L213-8 to l213-8-4 and stipulate that a drainage basin committee is created in each drainage basin or a group of drainage basins, referred to in Article L. 212-1. The committee consists of 40% of members of Parliament and senators and their deputies, representatives of the ministry and voivodeship assemblies, and, in a majority, representatives of communes or groups of local authorities competent in water management. Another 20% of the committee consists of representatives of non-economic water users, aquatic environments, marine environments, biodiversity, approved associations for environmental protection and consumer protection, and representatives of fisheries, as well as qualified personages. The third group, which also accounts for 20% of the committee, consists of economic users of waters, aquatic environments, marine environments and biodiversity, and professional organisations. The fourth and final group, which also accounts for 20% of the committee, consists of representatives of the state or its public institutions.[1] The chairperson of a committee is elected by representatives of the first three groups of which it is comprised. The drainage basin committee is consulted on the reasonability of taking significant actions that are the object of joint interest with respect to a predicted drainage basin and, more generally, on all issues addressed in Chapters I to VII of this title of the Code of Environment. It specifies the directions of the water agency's actions and participates in preparing the agency's financial decisions on conditions specified in Article L.213-9-1.[2] Members of four colleges representing a drainage basin can establish a territorial commission (*Fr. commission territoriale*), whose task is to propose priorities concerning the drainage basin committee's activities.

Additionally, Articles L213-8-1 states with respect to establishing water agencies that a water agency operates in a drainage basin or in a group of drainage basins, which is a state public administrative body (*Fr. établissement public de l'Etat à caractère administratif*). Such an agency deals with sustainable management of water resources and aquatic environments, drinking water supply, anti-flood regulation, and sustainable development of business activity. It can contribute to exploring, protecting, and preserving biodiversity on land and in the sea, as well as the marine environment in particular as part of following the national and regional strategies of biodiversity conservation. A water agency is managed by a board of directors, which comprises five groups of members. First, the coordinating prefect of the region where the agency is located and who chairs the board. Second, representatives appointed by individuals referred to in 1° of Article L. 213-8 within them.[3] Third, representatives appointed by individuals mentioned in

2° of Article L. 213-8 within them and representatives appointed by individuals mentioned in 2° bis of Article L. 213-8 within them. Fourth, representatives of the state or its public institutions, and fifth, a representative of the agency's personnel.

Another Article, L213-8-3, stipulates that the water agency board appoints an aid commission, which distributes financial aid granted by the water agency. The commission consists of representatives of various colleges present on the board of directors. Its deliberations and decisions are publicised. Additionally, Article L213-8-4 specifies that in order to prevent conflicts of interest, a member of the water agency board has to follow the ethics principles, and the board members make public declarations in this regard.

Project managing bodies (*Fr. organismes à vocation de maîtrise d'ouvrage*) are another group of institutions related to protecting water resources in France. They are divided into two groups: territorial public establishments for the river basin (*Fr. établissements publics territoriaux de bassin*) and marsh water management (*Fr. gestion de l'eau du marais poitevin*).

The first of this group of bodies, i.e., territorial public establishments for the river basin (*Fr. établissements publics territoriaux de bassin*), is regulated in Articles L213-12 of the French Code of Environment (Drobenko, 2008). These institutions, referred to as the EPTB, are French public institutions dealing with river and large river development and management within their geographical basin or drainage area. EPTB have a status of mixed consortia (*Fr. syndicats mixtes*) or inter-ministry institutions, and they are financed by territorial local governments, which are also their members. There were 32 such institutions operating in France in 2021. Territorial public establishments for the drainage basin have the necessary competences to act in three main areas. The first one is low water level management, flood prevention, and drinking water production. The second action affects the environment and includes actions for migrating fish and river bank maintenance. These actions are conducted within the area of local development, i.e., actions for the benefit of natural and cultural heritage. Each EPTB operates on a specific geographical territory, which often has an interdepartmental, inter-regional, or even international dimension. Therefore, it operates as a solidarity catalyst in a drainage basin or drainage area, and it participates actively in regional planning. The institution's operation is in line with the territorial solidarity rules, in particular with respect to areas of anti-flood expansion, which lie at the base of flood risk management.

A territorial public institution for the drainage basin can also establish – having consulted the committee for the drainage basin and local water commissions concerned, if such exists – a development project of common interest. It submits it to communes and public institutions with a view to intercommunal cooperation and to public institutions concerned, which deal with development and water management, which – if they approve it – authorise it to execute it. The resources of the public territorial facility include contributions from its members, subsidies, and loans, as well as funds collected by the water agency upon the facility's request, in accordance with Articles L. 213-10-9. Therefore, the funds of the public institution dealing with development and water management comprise contributions from its members, subsidies, and loans.

Water and biodiversity committees and water bureaus of overseas departments (*Fr. Comités de l'eau et de la biodiversité et offices de l'eau des départements d'outre-mer*) are another, fifth, institutional form of water management and protection in France. They are regulated in Articles L213-13 to L213-20 of the French Code of Environment.

Thus, Section I of Article L213-13 states that a water office (*Fr. offices de l'eau*) is established in each overseas department, which is a local public administrative institution joined to the department. In collaboration with the Committee for Water and Biodiversity (*Fr. Comités de l'eau et de la biodiversité*) and in accordance with the rules of managing natural resources and the environment laid down in Article L. 110-1, the water office is responsible for facilitating various actions of common interest in the field of water management and aquatic environments. Without prejudice to the competences passed on to the state and local authorities in this regard, it performs three main missions. The first one is to examine and monitor water resources and aquatic and coastal environments and their use. Another mission is to provide advice and technical assistance to ordering parties, training and information concerning water management and aquatic environments. The third mission is to programme and finance actions and work at the request of the Committee for Water and Biodiversity. Moreover, in accordance with international commitments of France and within the conventions subject to an opinion issued by the Committee for Water and Biodiversity, the water office can perform actions as part of international cooperation concerning water and sanitary issues within the limit of 1% of its funds, as applicable and in accordance with the applicable laws for each category of personnel, through the agency of its representatives.

Concerning water office organisation, Section II of Article L213-13 specifies that a water office is managed by the management board (*Fr. conseil d'administration*), which comprises representatives of the region, department, and communes, as well as public intercommunal cooperation facilities or mixed consortia with competences in water management, representatives of state services in the department, representatives of users and social and professional circles, representatives of approved associations of consumers and environmental protection, and individuals qualified in regard to aquatic and coastal environment. It is noteworthy that the members appointed as part of the first group (i.e., local government representatives) account for at least 50% of the management board and that the office is managed by the chairman of the board at the department level (*Fr. conseil départemental*). The prefect also performs the function of a government commissioner for the office.

Article L213-13, Section IV identifies the four main sources of funds at the disposal of water offices: fees mentioned in Article L. 213-14, 2, service fees, subsidies, and other financial resources provided for in applicable laws.[4]

The next article in this section of the Code of Environment, i.e., Article L213-13-1, specifies that the Committee for Water and Biodiversity (*Fr. Comités de l'eau et de la biodiversité*) in overseas departments consists of representatives of local authorities located in whole or in part within the drainage basin; representatives of users and individuals with special qualifications in this regard; and representatives of the state, of public institutions concerned, and of professional circles designated by the state. The committee is consulted on the reasonability of work and investments, which are the objects of joint interest planned for the drainage basin. The Committee for Water and Biodiversity performs missions in overseas departments entrusted to the Regional Committee for Biodiversity referred to in Article L. 371-3, which deals with informing, exchanging, and consulting on all issues related to biodiversity on land, in coastal areas, or in the sea, especially with regard to ecological continuity. It can be consulted on any issues that can have a significant impact on biodiversity. It also ensures the role and missions which are assigned to the committee for the drainage basin (*Fr. Comités de bassin*), specified in the code for the hydrographic basin in each overseas department.

## Legal and financial instruments in surface water protection

In France, according to Article 34 of the Constitution of 1958, the finances of local government units, as well as their powers, are determined by statute. However, the law must respect the principle of free administration of local authorities, according to which, under the conditions provided for by law, these communities govern themselves freely through elected councils and have regulatory power to exercise their powers. As a result, local governments use funds that they can freely dispose of under the conditions specified by law, and their resources consist of four main parts: local taxes, state grants and subsidies, advances and loans, and income from property and services provided. In this respect, Article 72-2 of the French Constitution specifies that in order to maintain the administrative freedom of local authorities, tax revenues and other resources of local authorities should constitute for each category of units a decisive part of their total resources. This decisive share of local taxes was further specified in the Organic Law of 29 July 2004 on the financial autonomy of local governments.

Finances in the field of water protection, regulated in Articles L213-9 to L213-9-3 of the Code of Environment, are an important aspect of the operation of the institutions, in particular water agencies. Thus, the financial resources of a water agency comprise, in particular, fees collected under Article L. 213-10 ff, repayment of advance payments made by it and subsidies paid by public bodies. Article L213-9-1 additionally specifies that a multi-year intervention programme for each water agency sets the areas and conditions of its operation for the mission determined in Article L. 213-8-1 and specifies the amount of expenses and income necessary to complete it. The Parliament determines the priorities of the multi-year intervention programme of water agencies and establishes the limit for their expenses during a period, as well as the amount of agency contributions paid to the French Office for Biodiversity (*fr. Office français de la biodiversité*). Meetings of a water agency management board regarding a multi-year intervention programme and licencing fees are convened upon the consent of the river basin committee, in accordance with the regulations concerning the multi-year amount of its spending and its distribution among the main intervention areas, which are the object of joint management by ministers for environment and finances, issued after an opinion of the National Water Committee is sought. These issues are additionally dealt with by the regulation contained in Article L213-9-2, which – following the 2019 amendments (*fr. loi n° 2019-773 du 24 juillet 2019 portant création de l'Office français de la biodiversité, modifiant les missions des fédérations des chasseurs et renforçant la police de l'environnement*) states that as part of its multi-year intervention programme, the water agency provides direct or indirect financial aid in the form of a subsidy, a performance bonus or refundable prepayments to public persons or individuals to enable them to execute actions or work which are the object of common interest, in the drainage basin or a group of basins, which contribute to sustainable management of water resources, aquatic environments, marine environment or biodiversity. Moreover, the agency participates financially in developing local plans and water resource management plans. Also, in accordance with the international commitments of France and under agreements subject to an opinion issued by the river basin committee, the agency can conduct actions as part of international cooperation concerning water and sanitary installations within the limit of 1% of its funds, as applicable, and in accordance with the applicable laws for each category of personnel, through the agency of its representatives. Upon request of a public drainage basin territorial unit, the water agency can collect fees established by this unit for the services provided under Article L.211-7. The whole revenue from the fees is paid to the budget of a territorial public facility of the drainage basin after deducting the management costs. The water agency also makes financial contributions to actions performed by the French Office for Biodiversity and grants capital subsidies to local governments and their groups for executing water supply and sanitary work in rural areas.

The fees (*fr. redevance*) that may be collected by water agencies include the following six contribution types. The first one is the fee for water pollution (*fr. redevances pour pollution de l'eau.*) regulated in Articles L213-10-1 to L213-10-4. The next one is a fee for modernisation of a water collection network (*fr. redevances pour modernisation des réseaux de collecte*) which is regulated in detail by Articles L213-10-5 to L213-10-7, and another – a fee for emitted pollution (*fr. redevances pour pollutions diffuses*) – is regulated in Article L213-10-8. The next one is a fee regulated in Article L213-10-9 for intake from water resources *(fr. redevances pour prélèvement sur la ressource en eau)*, and the fifth one, regulated in Article L213-10-10, for water retention when the water table level is low (*fr. redevance pour stockage d'eau en période d'étiage*). The sixth and final fee is one associated with the protection of the aquatic environment (*fr. redevance pour protection du milieu aquatique*), which is addressed in Article L213-10-12 of the Code of Environment.

These fees are consistent with the rule of preventing or remedying damage to the environment. The water agency establishes and collects fees for damage to water resources, marine environment, and biodiversity, in particular for polluting waters, intended for modernisation of sewage systems, for area pollution, for water intake from resources, for water retention during low tide, for obstacles in watercourses, and for protecting the aquatic environment.

A specific instrument within the ecological planning in France is the action plan for resilient and concerted water management whose aim is to guarantee quality water for all residents. This action plan has 53 measures, which aim to respond to three major challenges: (1) sobriety of use, (2) quality, and (3) availability of the resource. This plan also makes it possible to improve the response to drought crises.

The fight against water pollution is also related to this plan. The problem is represented by nitrates, which are very soluble in water and represent the main cause of pollution, which contributes to changing the biological balance of the aquatic environment. A majority of this pollution is caused by surpluses between nitrate inputs in the form of fertilisers and the rest, which are actually consumed by plants. Excess nitrates in the soil can leach out and pollute groundwater. Nitrates can also get into surface water through runoff. Water pollution by nitrates is one of the major causes of eutrophication of waters, particularly coastal waters.

For that reason, the European Directive 91/676/CEE called 'Nitrates' was introduced, which aims to reduce water pollution by nitrates of agricultural origin and eutrophication. This European directive requires Member States to define areas at risk of nitrate pollution, i.e., areas threatened by this pollution and areas suffering from eutrophication. In France, the criteria for vulnerable areas are defined by the order specifying the criteria and methods for evaluating the nitrate content of water (*fr. Arrêté n°0059 du 5 mars 2015 précisant les critères et méthodes d'évaluation de la teneur en nitrates des eaux et de caractérisation de l'enrichissement de l'eau en composés azotés susceptibles de provoquer une eutrophisation et les modalités de désignation et de délimitation des zones vulnérables définies aux articles R. 211-75, R. 211-76 et R. 211-77 du code de l'environnement.*).

Vulnerable areas are designated by the basin coordinator prefects (*fr. les préfets coordonnateurs de bassin*). These areas are revised at least every four years, based on the results of monitoring campaigns and an institutional consultation phase. The 'Nitrates' Directive specifies that an action programme must be made obligatory in vulnerable areas and sets certain provisions. In France, the action programme is composed of the national action programme (*fr. programme d'actions national*) – i.e., national base applicable in all vulnerable areas and regional action programmes (*fr. programmes d'actions régionaux*) which reinforce and complement certain provisions of the national action programme.

The national action programme defines the measures to be implemented to combat water pollution by nitrates. In accordance with the Nitrates Directive, the action programme must be revised every four years. The 7th National Action Programme was published on 10 February 2023, and its measures are valid from 1 January 2024 (these measures are supplemented and reinforced by valid regional programmes). Two bodies contribute to the development and monitoring of the implementation of this public policy – (1) the Nutrient and Environmental Emissions Management Consultative Group (*fr. le groupe de concertation gestion des éléments nutritifs et des émissions vers les milieux*) composed of stakeholders in the fight against nitrate pollution whose task it is to monitor the progress of work related to the development, implementation, and evaluation of the technical and regulatory framework in the application of the Nitrate Directive and public policies related to the management of nutrients and related emissions to the environment and (2) the Scientific and Technical Committee for Nutrient and Environmental Emissions Management (*fr. le comité scientifique et technique gestion des éléments nutritifs et des émissions vers les milieux*) composed of experts appointed *intuitu personae*. Its aim is to inform the public administration and the advisory group by issuing opinions and recommendations.

The general tax on polluting activities, which was introduced in 2000 in France, can also be mentioned as a partly financial, partly legal instrument. The aim of this tax is to promote the reduction of phosphates in detergents and also in fabric softeners, which ultimately pollute water. Phosphates in textile detergents intended for consumers were banned from 1 July 2007 in France. In combination with the general tax on polluting activities, this action has the effect of reducing the phenomenon of eutrophication by 20% to 25%. Finally, in 2012, the European Regulation on Detergents CE No. 648/2004 was revised. For the first time at the European Union (EU) level, measures are being introduced to limit phosphates and phosphorus compounds for consumer detergents.

In connection with water pollution, it is also necessary to reflect on the issue of pesticides that are contained in various preparations. Weed preparations used on impervious or poorly permeable surfaces (e.g., gravel yards or garage slopes) end up in surface or groundwater and cause run-off-related water pollution due to low infiltration. For that reason, France adopted Act No. 2014-110 of 6 February 2014, amended in 2015. This act imposed from 1 January 2017 the goal of eliminating pesticides in all public spaces. As of 1 January 2019, the use of these same products by individuals has been prohibited.

**Role of organisation and of society in surface water protection**

Historically, there is a very important role of the associations in the French legal system. The old regulation contained in Article 1 of the law of 1901 defined the association as the agreement by which two or more people pool and share, on a permanent basis, their knowledge or their activity for a purpose other than sharing profits. Thus, three elements that constitute the association are contractual agreement, share of knowledge or activity, and selfless purpose.

Participation of the society in water protection was regulated widely in the French Code of the Environment because it is dealt with in the whole Title I of Book I, i.e., Articles L110 to L191. Given the highly complex method of regulation, which cannot be discussed in detail because of the size of this study, one should only point out that sharing information about the environment and its protection and the society's participation in it and in environmental impact assessments is very broad in France, with quite specific regulations regarding the nuclear industry – for which water for reactor cooling is essential in France. Thus, a special chapter was devoted to the right to information on nuclear issues (Articles L125-10 to L125-11) and one concerning transparency in nuclear issues (Articles L125-12 to L125-16-1).

It is noteworthy that a major role in France in this regard is played by associations for environmental protection (*fr. Associations de protection de l'environnement*) and local government units, which are dealt with in two chapters of the code – Articles L141-1 to 141-2 and from L142-1 to L142-4, respectively. It is also noteworthy that according to Article L141, associations for environmental protection gain subjectivity regarding consultations and a mandate for being a party to court and administrative proceedings only if they have been conducting their activities for at least three years, i.e., they pursue their statutory activities within the scope of environmental protection and management of wild fauna and flora, improvement of the life environment, and protection of waters, air, soil, places, and landscapes, and they have been granted the status of approved environmental protection associations (*fr. associations agréées de protection de l'environnement*). This status is granted on the conditions defined by the decree of the Supreme Administrative Court – *Conseil d'Etat*. A decision to grant this status is valid for an indefinite time and within a specific framework, considering the territory in which the association actually conducts its activities. The status of an approved association for environmental protection is renewable, but it can also be rescinded when the association no longer meets the conditions it met upon being granted it.

Moreover, the French legislation regulated not only local government units' participation in proceedings and settlements associated with environmental protection in Articles 142-1 to 142-4 but also the rules for the approved environmental protection associations entering ongoing proceedings. This regulation, which is *lex specialis* for the general rules, was necessary not only because of the large number and popularity of associations in France (which are a very important form of citizens' participation in public life) but also because France is the country with the largest number of local government units at the commune level.

When it comes to the general rules of public participation in making decisions concerning environmental protection, including water protection, Articles L110-1 to L110-7, as well as L120-1, deserve attention. The last of them emphasises that society participates in developing public decisions with an environmental impact in order to improve the quality of the public decision-making process and to provide its democratic legitimisation, preserving a healthy environment for the present and future generations, as well as improvement and diversification of information on the environment. Section II of this article specifies that such participation gives society a number of rights, with the major ones being the right of access to relevant information that enables effective participation, the right to demand a participation procedure on conditions provided for in this code, the requirement of a reasonable time limit for putting forward suggestions and proposals, and the right to receive information on the way in which their comments and proposals are taken into consideration in a permit or approval.

When it comes to specific rules of the society's participation in making decisions on environmental protection, there is a chapter in the Code of Environment on the society's participation in developing plans, programmes, and projects with environmental impact (i.e., from Articles L121-1-A to L121-24). Another code fragment includes Articles L122-1 to L122-15, which make up a chapter on environmental impact assessments (*fr. evaluation environnementale*). The third separate fragment of the code concerns public participation in making decisions with environmental impact contained in Articles L123-1-A to L123-19-11. These regulations are expanded in Articles L123-20 to L123-33, which were grouped in the Code part entitled 'Local Consultations on Projects with a Potential Environmental Impact' – which is another indication of a highly complex regulatory sphere in this regard with respect to the local government units. Articles L124-1 to L124-8 make up another specific chapter entitled 'Right to Access Information on the Environment'. Other separate chapters include Chapter V, entitled 'Other Methods of Passing Information' (Articles L125-1 to L125-40); Chapter VI, entitled 'Project Declarations' (Article L126-1); and Chapter VII, entitled 'Infrastructure of Geographical Information' (Articles L127-1 to L127-10)

To conclude, like other European legal systems, French law contains several instruments for surface water protection. They are associated with planning (state and local plans, water management programmes), administration (system of permits, opinions, agreements), information (monitoring, information for society and public organisations), and special issues (associated with the nuclear industry).

**Penal responsibility**

As for penal reasonability, it is worth noting that the currently applicable new penal code *(fr. le nouveau code pénal*) was established by a number of laws of 22 July 1992, thus replacing the penal code of 1810, which came to be referred to as the old penal code (Kolb & Leturmy, 2022). Ultimately, when it fully entered into force on 1 March 1994, the new French Penal Code was considered a modern codification containing many pioneering solutions (Mariański, 2023). One of them was the introduction of criminal liability of collective entities and the inclusion of many regulations existing so far in special acts into one legal act while marginalising the importance of extra-code criminal law regulations (Rogacka-Rzewnicka, 2004).

However, this process of marginalising the importance of extra-code criminal law regulations was not applicable to environmental protection regulations, as in this field, we have to refer to the French Code of Environment and not to the French Penal Code. As a result, a special legal branch of law called – environmental penal law (*fr. droit pénal de l'environnement*) was developed in France. Environmental criminal law is defined as the branch of law criminal law which prevents and punishes attacks on the environment. This is a right annexed to all the provisions concerning environmental law, bringing together a large number of offences with different characteristics. It presents two orientations:

- A disciplinary orientation of a technical nature due to the accumulation of special policies. This is demonstrated, for example, by the incriminations intended to repress non-compliance with the requirements provided for by the legislation on classified installations.
- A sanctioning orientation of a moral nature. It reflects the recognition of the environment as a value to be protected. This is the case, for example, with incriminations relating to freshwater pollution (Beziz-Ayache, 2022)

In the Code of Environment, in Book II (*Livre II*), which contains a separate Title II *(Titre II)*, a separate fragment in Chapter VI (*Chapitre VI*) is devoted to sanctions and control.

It was pointed out in the code fragment in question, with respect to administrative sanctions in Article L216-1, that if this title were to be applied, a call for remediation of a defect issued under Articles L. 171-7 and L. 171-8 could provide for having to perform all the additional controls or analyses, whose costs will be borne by the operator or the owner (whose activities are being controlled). Additionally, in order to apply this title, the *ex officio* administrative execution measures taken as per L. 171-8 sect. 2 II can be entrusted to entities mentioned in Article L. 211-7-1 (i.e., local government units or water agencies) with their consent.

With respect to penal sanctions, Article L216-3 specifies at the beginning that apart from court police officers and agents and environmental protection inspectors referred to in Article L. 172-1, additional entities are authorised to investigate and register violations of provisions of Chapters I–VII of this title, as well as the texts and the decisions to implement them. These additional entities include representatives of state services, personnel of state forests, nuclear security inspectors, researchers, engineers, and sworn technicians of the French Marine Exploitation Research Institute; harbour officers and their assistants; rural guards; customs officers; coast guards; and nature reserve representatives.

Another important rule in this regard is stipulated by Article L216-4, according to which all the costs of controls, expert opinions, and analyses are borne by the operator or, if there is none, the owner. Additionally, as per Article L216-5, if a crime results in the destruction of fish spawning, growth, or grazing areas, or it disrupts ecological continuity or the minimum flow of a watercourse, a copy of the protocol is sent for information to other bodies competent for the aquatic environment protection.

A detailed description of different types of crimes can be found in subsequent articles of the Code of Environmental Protection. Thus, Article L216-6 reads that discharging, pouring, or allowing the run-off to surface, ground, or marine waters within the territorial waters, directly or indirectly, of any substance(s) whose effect or reaction has – if only temporary – harmful impact on human health or on flora or fauna, except for the damage referred to in Articles L. 218-73 and L. 432-2, or considerable changes in water supply or restriction in bathing area use, is punishable by a prison sentence of up to two years and a fine of 75,000 euros.[5] If the wastewater removal/discharging operation is allowed under a relevant permit, the provisions of this section are applicable only when the conditions laid down in the permit are not met. A court may also obligate a convicted entity to restore the status of the aquatic environment in accordance with Article L. 173-9. The same punishments and measures are applied for discarding or leaving considerable amounts of waste in surface or groundwaters or in marine waters within the territorial waters, on beaches, or on the sea coast. This does not apply to discharge from ships at sea. The statute of limitations for public actions related to the felonies mentioned in this article runs from the date when the damage was found.

Another felony, regulated in Article L216-7, involves unlawful exploitation of a structure. Thus, exploitation of a structure without observing the regulations concerning, in particular, migrations of fish, the minimum water flow, or having to observe the easement for the public as provided for in Article L214-9 is punishable by a fine of 75,000 euro.

Article L.216-13 is the last penal provision within the scope in question. It stipulates that if the conditions laid down in Articles L. 181-12, L. 211-2, L. 211-3, L. 214-1 to L. 214-6 or the measures introduced under Article L. 171-7 of the code or Article L. 111-13 of the mining code are not observed, a judge may, on a petition from a prosecutor, acting *ex officio* or on a petition from an administrative organ, a harmed party or a registered environmental protection association, impose all the necessary and useful remedies on individuals and legal persons for the maximum period of one year, including, in particular, suspension of or a ban on the illegal activity.

## Summary

The French legal system is characterised not only by a quite original catalogue of sources of law, which strongly emphasises the role of the courts, but also by a quite specific system of tax law and financial law. French law may also be a very interesting comparative element of the analysis because of historical and methodological factors. In the past, many European legal systems were often inspired by French regulations, mainly in the field of civil regulations. That is why an assessment of the legal solutions concerning water management and protection of waters in French law may contain several basic observations that may be useful in a broader comparative analysis.

First, it should be regarded as a very positive fact that – as a rule – most of the aspects of water management and river water protection are laid down in a single, very complex legal act, namely, the Code of the Environment (*Fr. Code de l'environnement*). Title I of Book II was isolated as part of the code, entitled 'Waters and Aquatic and Marine Environment', which corresponds to Articles L210-1 to L219-18 of the said legal act, to which provisions of the general code of local governments (*Code général des collectivités territoriales*) apply only incidentally by reference.

Second, an important role is played not only by local government units with which the institutional aspect is linked but also by a prefect who – to put it simply – performs a function similar to that of a voivodeship governor, which is known in Polish law.

It is a typical feature of the French law that there are so many institutions. This should be regarded as a negative feature, as it makes the system complicated and may breed competence disputes. There are ten types of institutions of different kinds:

1. National Water Committee (fr. Comité national de l'eau)
2. Prefect coordinator for water bodies (Fr. Préfet coordonnateur de bassin)
3. Committees for drainage basins (Fr. Comités de bassin)
4. Water agencies (Agences de l'eau)
5. Project managing bodies (Fr. organismes à vocation de maîtrise d'ouvrage) as territorial public establishments for the drainage basin (fr. établissements publics territoriaux de bassin)
6. Project managing bodies (Fr. organismes à vocation de maîtrise d'ouvrage) as marshy waters management (Fr. gestion de l'eau du marais poitevin)
7. Committees for water and biodiversity (Fr. Comités de l'eau et de la biodiversité)
8. Water offices of overseas departments (Fr. offices de l'eau des départements d'outre-mer)

9 National Committee of the French Initiative for Coral Reefs (Fr. Comité national de l'initiative française pour les récifs coralliens)
10 Permanent Technical Committee for Dams and Hydrostructures (Fr. Comité technique permanent des barrages et des ouvrages hydrauliques)

Fourth, the role of case law, including *Conseil d'Etat's* power to issue decree laws, allows for the creation of rather general regulations, which are then made more specific by case law.

Fifth, the importance of water in France's internal policy is very high also because of the dominant share of nuclear energy in its energy mix, with nuclear reactors being cooled down with water.

Sixth, the scope of general competence and implementation of the water directive in France is similar to other EU countries. The French legal system provides for a multitude of institutions in this regard.

Finally, the challenges facing water protection and coordination of this policy with local government units have undoubtedly become more important during the coronavirus pandemic and in the wake of hostilities in Ukraine. While these challenges have slightly different dynamics and structures in each EU Member State, the description of water protection in France may be an interesting inspiration for further comparative research in this area. It will also be important to ultimately optimise the administrative and procedural costs of water protection, which can be achieved both by computerisation and by sharing them within the framework of cooperation between several local government units. Coordinated activities related to legal and technological changes should be carried out in a way that ensures a safe level of revenue for the institutions that are responsible for water protection, especially at the local level. Therefore, this issue should be the subject of a long-term strategy for development in this field, which should respect the principle of legal certainty from both the perspective of the authorities and users.

## Notes

1 When an organization is called for nominating several representatives to the basin committee, the nominations are made in such a way as to make the difference between the number of nominated males and the number of nominated females not greater than one.
2 To this end, state representatives in each department in the drainage basin submit to the drainage basin committee the state priorities and major projects for each of the departments concerned once every three years.
3 That is, an MP and senator and, for each of them, a deputy with the MP or senator status, representatives of regional and voivodeship assemblies, and, in majority, representatives of communes associations of competent bodies dealing with water.
4 Control of legality and budget control of the office acts is performed in accordance with Article L. 3241-1 of the general code of local government units (*Code général des collectivités territoriales*).
5 It should be emphasized that the 1992 penal code now valid in France defines only the maximum punishment, below which a punishment is established at the judge's discretion.

## References

Beziz-Ayache, A. (2022). *Fiches de droit pénal de l'environnement*. Ellipses.
Blanchon, D. (2024). *Géopolitique de l'eau Entre conflits et coopérations*. Le Cavalier Bleu.
Chapus, R. (1999). *Droit du contentieux administrative*. Montchrestien.
Defeuilley, C. (2024). *La politique publique de l'eau: Gouverner un bien commun*. Le Bord de l'eau.
Drobenko, B. (2008). *L'essentiel du droit de l'eau: À jour de la loi sur l'eau et les milieux aquatiques et de ses décrets d'application*. Gualino.
Fabre-Magnan, M. (2014). *Introduction au droit*. Presses Universitaires de France-PUF.
Klimaszewska, et al. (2017). *Przepisy o spółce z ograniczoną odpowiedzialnością (societé à responsabilité limitée) w francuskim Kodeksie handlowym (Code de commerce). Tekst przepisów wraz z tłumaczeniem oraz porównaniem do polskiego Kodeksu Spółek Handlowych*. Wydawnictwo UWM.
Kolb, P., and L. Leturmy (2022). *Cours de droit pénal general*. Gualino.
Mariański, M. (2021). Formy realizacji zadań własnych przez jednostki samorządu terytorialnego we Francji ze szczególnym uwzględnieniem spółek handlowych. *Prawo i Więź* 1 (35):128–139. https://doi.org/10.36128/priw.vi35.249
Mariański, M. (2023). Wprowadzenie do problematyki szpiegostwa we francuskim kodeksie karnym. *Opolskie Studia Administracyjno-Prawne* 21 (2):197–209. https://doi.org/10.25167/osap.5164
Martin, G., and J.-L. Gaffard (2023). *Droit et économie de la transition écologique: regards croisés*. Mare&Martin.
Prieur, M. (2023). *Droit de l'environnement*. Dalloz.
Roche, C. (2022). *L'essentiel du droit de l'environnement*. Gualino.
Rogacka-Rzewnicka, M. (2004). Prawo karne. In *Prawo Francuskie. Tom I*, edited by A. Machowska and K. Wojtyczek, 449–496. Zakamycze.

# 4 Legal protection of surface waters in Spain

*Jarosław Dobkowski and María José Cazorla González*

## Introduction

Water has been the subject of European regulation since the 1970s, but the Union's legislative process itself shows changes in the way this natural resource is managed and protected. Specifically, water policy and the legal provisions for the conservation of water resources have been reinforced as environmental policy has been strengthened in European law, covering more and more realities (rivers, seas, lagoons, agriculture, urban supply, etc.); the water policy has been strengthened as a result of the increasing number of water-related issues (rivers, seas, lagoons, agriculture, urban supply, etc.).

One can claim with respect to the hydrological picture of Spain that water shortage has always been felt there. Certain legal institutions – some of them even of Arab origin – were developed there (Martín-Retortillo Baquer, 1960, 11–54; Schwarz, 1966; Organización de las Naciones Unidas para la Agricultura y la Alimentación, 1976, 26–29; Vergara Blanco, 1991, 137–161). The first laws concerning water were passed as far back as 1866 (Ley de Aguas, de 3 de Agosto de 1866) and 1879 (Ley de Aguas, de 13 de junio de 1879) (Guaita Martorell, 1982; Gallego Anabitarte et al., 1986; González Pérez et al., 1987). The landmark was the Water Act No. 29/1985 (Embid Irujo, 2009, 290–294). Because of the drought of recent years, some extraordinary measures have been implemented, which – after subsequent renewals – are becoming permanent. The Water Framework Directive (WFD) was implemented in Spain by the Royal Law Decree 1/2001 of 20 July 2001 containing the Water Law, as well as a later act 62/2003 of 30 December 2003 on fiscal, administrative, and social measures, whose Article 129 introduced quite important changes in the said Water Law. The amended act 10/2001 of 5 June 2001 of the National Water Plan by the act 11/2005 is also important. These acts have established the domestic legal framework for water policy for the competent bodies and institutions (Menéndez Rexach, 2010, 23–67; Agudo González, 2010, 69–141).

The Water Law, in its formal and legal sense, denotes the Royal Decree Law 1/2001 of 20 July 2001, which was based on the existing Water Act 29/1985 of 2 August 1985. Given the position of royal law decrees in the

DOI: 10.4324/9781003586913-5

Spanish legal system, what took place was not so much a technical incorporation of the valid regulations but the issuing of a new normative act (Sánchez Camacho, 2007, 58–78) as well as the acts referred to above.

The national body of the Spanish law concerning water policy also comprises other normative acts, in particular, the Royal Law Decree 11/1995 of 28 December 1995, which establishes the rules applicable in urban wastewater treatment together with the Royal Decree 509/1996 of 15 March of 1996, which expands on it, as well as other royal decrees: the Royal Decree 849/1986 of 11 April 1986 on the public water domain, amended the Royal Decree 907/2007 of 6 July 2007, approving the royal decree on water planning (907/2007, 1159/2021), the Royal Decree 1620/2007 of 7 December 2007, which establishes the legal regime for reuse of treated water, Royal Decree 60/2011 of 21 January 2011 on the environment quality standards with respect to water policy and the Royal Decree 817/2015 of 11 September 2015, establishing the criteria for monitoring and evaluation of surface waters and the environmental quality standards. To a certain extent, the Royal Law Decree 1/2016 of 16 December 2016, approving the consolidated text of the law on integrated pollution prevention and control also applies.

Pursuant to Article 1, the aim of the Water Law is to regulate the public water domain, normalising the rules of water use and exercising the rights granted to the state in matters related to that domain as part of the state competence laid down in Article 149 of the Constitution of Spain of 27 December 1978. The Water Law also establishes the basic rules of inland, coastal, and transitional water protection without detriment to their legal classification and specific regulations applicable to them. In this regard, inland surface waters, including renewable groundwater integrated with the hydrological cycle, constitute a uniform resource subordinate to the general interest, which is part of the national, general social domain as a public water domain. The state is responsible for water planning, covering all the actions on public waters in each case and on conditions laid down in these regulations (Chiu, 2014).

However, the Spanish legislation is applied under the interpretation criteria and guidelines established by the European Union (EU), which influence the application of all these regulations: the European Green Deal, both the Circular Economy Action Plan and the new EU Climate Change Adaptation Strategy refer to a wider use of reclaimed water as a way of increasing the EU's capacity to respond to the growing pressures on water resources (Molina Giménez, 2022, 311–343).

Under Regulation (EU) 2020/741 of the European Parliament and of the Council on minimum requirements for the reuse of water, the urban wastewater treated in accordance with the requirements of Directive 91/271/EEC on urban wastewater treatment (Urban Waste Water Treatment Directive) must undergo additional treatment to meet new minimum quality parameters and be suitable for use in agriculture (Molina Giménez, 2021, 147–202).

The proposed revision of the Urban Water Treatment Directive strengthens the existing provision encouraging water reuse by requiring Member States to systematically promote the reuse of reclaimed water from all urban wastewater treatment plants. A swift adoption of the proposal, with its requirements for improved surveillance, monitoring, and reduction of pollution at source, will improve the quality of reclaimed water, which will further facilitate its reuse (Hernández López & González Martínez, 2023, 231–248).

Water reuse also contributes to the Farm to Fork strategy's objective of reducing the environmental footprint of the EU food system and strengthening its resilience by providing an alternative and more reliable water source for irrigation. Under the Common Agricultural Policy, there are funding opportunities for irrigation investments with reclaimed water as an alternative water supply.

This new regulation is also in line with the Water Action Agenda adopted at the UN Water Conference held a few months ago in New York, with a particular focus on access to sufficient and good quality water and adaptation to climate change.

The aim is that this new regulation will increase consumer and farmer confidence in this circular approach to water use, ensure water security and reduce pressures on increasingly scarce water resources, promote the safe use of reclaimed water in agriculture, and ensure the protection of human and animal health and the environment and will be a tool to help protect biodiversity.

## Legal basis of surface water management and protection

In Spain, as in most of the surrounding countries, when talking about water management, a fundamental distinction is made between groundwater and surface water. The use of groundwater is a very old practice in our country, and its development has been linked to the evolution of technical, social, and economic factors, and is mainly linked to the private and family sphere, with little participation in the public sector. So the benefits and the negative effects, there is a need for specific groundwater regulation and controlled and optimal groundwater management. In this respect, it should be remembered that Article 45 of the current Constitution states that the public authorities must ensure the conservation of natural resources, but also that everyone has the duty to conserve them. Therefore, the legislator (national and regional) must provide incentives with the appropriate measures for the rational use of natural resources, for example, by ordering the exploitation of natural resources in a sustainable way (Flores García & Molina García, 2017).

The following comprise the public water domain: (a) inland waters, both surface and groundwater, renewable, regardless of the time of their renewal; (b) beds of natural watercourses, permanent or temporary; (c) bottoms of lakes and lagoons and the bottoms of surface water bodies in public watercourses; (d) aquifers, for the purposes associated with development of water resources; (e) water obtained by desalination of seawater (Article 2) (Döpp, 2000).

This does not apply to temporary watercourses or water bodies within private real estate, and their course or water quality changes must not be subject to the owner's actions (Article 5). This also does not apply to the ponds on private land, which is regarded as its integral part unless it is intended for the exclusive use of the owner of such land and it is without detriment to the environmental protection (Article 9).

This is important because – as a rule – individuals and other entities cannot become water owners, but they can only acquire the right to use them, except for the rights acquired under previous regulations, which regarded groundwaters as the property of the owner of the real estate where it was situated. Currently, the use of a water resource requires a previous concession as part of the state monopoly or a permit issued by the competent administrative authority. There are also water registers (Article 80), obligatory irrigation communities, and gathering water users (Article 81) (Fanlo Loras, 2006, 77–82; Bunclark et al., 2011, 48–55; Embid Irujo, 2017, 45–62; Wawer, 2021, 202).

Spain did not take over strict and exclusive water rationing, but – while observing the traditional rules of joint use of such resources by individual farmers – it created the legal framework for irrigation communities within the public law while respecting the classic forms of private law and mainly the local customs. A system of irrigation communities has been operating effectively for centuries. However, nowadays, groundwater has been exploited uncontrollably (Flores García & Molina García, 2017), which has resulted in a situation in which more water is collected illegally than from legal sources, which results in a crisis of the existing system solutions.

Nevertheless, Article 54 of the Water Law provides for the private use of public waters directly under the legal provision, whose beneficiaries include farm owners who can benefit from the rainfall flowing through their farms and standing in them without limitations other than specified in the act and arising from respecting the third-party rights, but with the general prohibition of abusing one's rights, and the owners of other real estates within the scope of using water from the sources situated in their area and groundwater if the annual volume does not exceed 7,000 $m^3$ unless the aquifers are officially regarded as exploited excessively or threatened with excessive exploitation (Maestre Rosa, 1969; Supreme Court, Rec. 2810/2008).

So the recognition of the public water domain in Spain is the basis for the status of individuals in their legal relationship with water. Therefore, it is on this basis that the legal regime of their rights and obligations is established. However, we are far from having a current and effective legal configuration of the right to use surface and groundwater because even in the 21st century, there are still doctrinal and jurisprudential discussions about the rights that individuals have with respect to water, its content, and its protection. It may be necessary, from a doctrinal and legal point of view, to configure the right to use water, regardless of the title that individuals provide for its exercise. In this way, all the uses made of water would be unified, regardless of when the right was acquired (Alcaín Martínez, 2010).

This is so in Spain because, as the Constitutional Court recognised, 'it is necessary to note that, without prejudice to their legal classification as waters of private domain', the legislation prior to the new Water Law did not establish a right of ownership over them that could be traced back to the general regime defined in Article 348 (La Ley 1/1889) of the Civil Code and in the concordant precepts. Private ownership of certain land waters was already in that legislation a 'special property' (Title IV of Book Two of the Civil Code), subject to strict limits as regards the powers of the owner (...), its content being limited to 'the use of the waters' (Tribunal Constitucional, Rec. 824/1985).

On the other hand, there are civil legal institutions: waterway easement (Article 47 ff), private and general water use, and common and specific (Article 50 ff). Interesting features in this regard include the assignment of the right to use water (Article 69) or accounting for the right to use water (Article 71).

Water management has been planned and directed in Spain for many years. The general objectives of water planning include achieving good public water status and proper protection, which also includes other waters covered by the Water Law; satisfying demand for water, sustainability, and harmonisation of regional and sectoral development; improving the resource availability; and protecting their quality, frugality, and rationalisation in their use in harmony with the environment and other natural resources. The water policy helps to implement sectoral strategies and specific plans developed by the public administration without detriment to the rational and sustainable resource management, which must be applied by the Ministry of Environment or by competent water administration authorities, which is a condition for further permits, concessions, and material assets which can be applied for. Plans are developed with the use of the National Water Plan and water plans for drainage basins (as per Article 16: an area of land whose surface run-off flows through a number of streams, rivers, and/or lakes to the sea through a single estuary or delta) based on the Water Law, whose territorial range coincides with that of the respective drainage basin organisation (Article 40). A drainage basin organisation is a resource management unit, in particular, an entity responsible for water planning and management, and it is appointed when a drainage basin exceeds the territorial range of an autonomous community, or it is appointed as an intra-community organ of water administration, which is also specialised, decentralised, and participative. The scope of activities of a drainage basin organisation of each type is specified in the Water Law (Article 21). This is important because the procedure for the preparation and review of water management plans in drainage basins is regulated by law, and schedules, work programmes, elements that require consideration, and preliminary projects are announced each time in order to inform the public and ensure consultations from the beginning of the preparatory process. Certain limitations arise from the Spanish-Portuguese convention of Albufeira of 1998, which defines the framework of bilateral relations concerning water resources. Programmes related to coastal and transitional waters are developed by the state administration or by autonomous communities which have a coastal line on their territory and take part in the work of a relevant committee. Water plans are developed in coordination with various sectoral plans, both with respect to water and land use, especially to the regulations concerning plans for irrigation and other agricultural use. Water plans are approved by state administration bodies (Article 41).

According to Article 45, in each case, the National Water Plan provides necessary measures for the coordination of various hydrological plans for a drainage basin, alternative solutions, a forecast, and conditions for water resources transfer between territories covered by various water management in a drainage basin, modification provided for in planning regarding the resource use for water supply to the population or irrigation. The valid National Water Plan was approved by the act 10/2001 of 5 June 2001 on the National Water Plan. Specific conditions of water planning are laid down in the amended Royal Decree 907/2007 of 6 July 2007.

Water planning makes it possible to sort out water management. The plans are not just simple programmes of work but normative instruments which adapt general regulations of the law to specific local conditions. Although water plans are key solutions of the Water Law, their development process is time-consuming, and there were delays in their approval process. Water plans must be taken into consideration in local plans.

The Water Law also regulates the issues related to public water protection and water quality. The statutory objectives of water protection, including public waters, as per Article 92, are as follows: (a) preventing deterioration, protection, and improvement of the status of aquatic ecosystems, as well as land ecosystems and wetland, which are directly dependent on aquatic ecosystems with respect to their water needs; (b) promoting sustainable use of water by protecting the available water resources and ensuring their sufficient supply in good status; (c) protection and improvement of the status of the aquatic environment by implementing specific measures in order to reduce gradually the discharges, emissions, and loss of priority hazardous substances; (d) gradual reduction of groundwater pollution and preventing its further pollution; (e) alleviating the effects of floods and droughts; (f) achieving, through implementation of the relevant regulations, of the objectives set in international treaties on preventing and eliminating marine environment pollution; (g) avoiding accumulation of toxic or hazardous compounds in the subsoil or other accumulations which could bring about degradation of public waters; (h) guaranteeing allocation of the best quality waters in an area or region for supply to the population.

The Water Law, in its Article 92 *bis*, also lays down the environmental objectives which serve water protection, including surface waters (prevention of the deterioration of the status of surface water bodies; protection, improvement, and restoration of all water bodies in order to achieve a good status of surface waters; reduction of pollution with priority substances; and elimination or gradual elimination of discharges, emissions, or loss of priority hazardous substances) for groundwaters (prevention or reduction of polluting groundwaters or prevention deterioration of the quality of all groundwater bodies; protection, improvement, and restoration of groundwater bodies; and ensure a balance between water intake and supply in order to achieve a good status of groundwater, reversing any considerable and permanent trends of growth of pollutant concentrations caused by human activity in order to gradually reduce the groundwater pollution), for protected areas (meet the requirements for protection standards valid in an area and achieving the environmental objectives specified in them), and for artificial and heavily modified water bodies (protection and improvement of the status of artificial and heavily modified water bodies in order to achieve a good ecological potential and good chemical status of surface waters). If water bodies are heavily modified as a result of human activities or if these objectives are unattainable or attaining them would entail high costs due to their natural conditions, less strict environmental objectives are set on conditions which will be defined in each case in water management plans.

Apart from water plans, water protection measures include various states and potentials, and the defined ones include surface waters, groundwaters, and artificial and heavily modified water bodies, with the technical conditions and criteria of their classification laid down in the royal decree. Moreover, there are water status monitoring programmes in each drainage basin in order to ensure its comprehensive and consistent assessment. The action programmes should aim at achieving environmental objectives. Such actions can be basic and supplementary, with the basic measures being the minimum requirements that must be met in each separation, and they are established by means of a regulation, with the supplementary measures being those which must be applied additionally to achieve the environmental objectives or additional water protection (Article 92. ter – Article 92 quáter). The specific rules in this regard arise from the Royal Decree 60/2011 of 21 January 2011 and the Royal Decree 817/2015 of 11 September 2015.

The Water Law also defines the concept of pollution in Article 93. This is understood to denote the action and effect of introducing materials or forms of energy or inducing conditions in water, which directly or indirectly imply a harmful change of its quality with respect to later applications, human health, or aquatic ecosystems or land ecosystems directly linked to aquatic ones, cause damage to property or deteriorate or hinder the use of the environment. Degradation of public waters covers adverse changes in the environment that affect water quality.

Article 97 of the Water Law forbids actions that can cause pollution but also degradation of public waters, in particular, (a) accumulating solid waste, rubble, or substances, regardless of their character or places of storage, which pose or may pose a threat of polluting waters or degradation of their environment; (b) performing actions on the physical or biological environment which is affected by water, which results or may result in its degradation; and (c) performing actions within the scope of protection, established within the hydrological plans, if they can pose a threat of pollution or degradation of public waters.

Moreover, according to Article 98, drainage basin organisations take the necessary measures when granting licenses and permits so that the use respect the environment and ensure the ecological flow or ecological requirements set in hydrological plans. When concessions and permits with an impact on public water areas which can entail a risk to the environment are considered, a report should be submitted on possible adverse impact on the environment, which will be forwarded to the authorities competent on the matters of the environment so that they can make a decision to implement corrective measures, which should be implemented as a result of such a report. If a drainage basin organisation assumes the existence of a serious threat to the environment, it also submits its position concerning the initiation of a procedure aimed at assessing the environmental impact for consideration by a competent environmental protection authority.

However, pursuant to Article 96, a drainage basin organisation can provide for a service area around a surface water body, which is necessary for its use, provided that, in each case, coastal areas of lakes, lagoons, and water bodies are subject to easements and protected zones. Therefore, there is a special area of an ecological nature in the form of a protected zone. Apart from protected zones, there are protected areas that are subject to protection under specific regulations concerning the protection of surface water or groundwater or the protection of habitats and water-dependent species. According to Article 99 *bis*, for each drainage basin there is at least one register of areas, which must contain (a) areas in which water is taken for human consumption, provided it ensures at least 10 $m^3$ daily or supplies more than 50 people, as well as outlined boundaries of protection; (b) areas which – according to the relevant water plan – are to be used for intake of water for human consumption; (c) areas notified for protection of aquatic species of economic importance; (d) water bodies intended for leisure, including those designated as bathing grounds; (e) areas defined as sensitive in accordance with standards of water protection against pollution caused by nitrates from agriculture; (f) areas defined as sensitive according to urban wastewater treatment; (g) areas designated for protection of habitats or species, in which maintaining or improving the water quality is an important factor of their protection; and (h) boundaries of mineral and thermal water protection, approved in accordance with specific.

We must remember that a quarter of a century ago, Royal Decree 261/1996 of 16 February 1996 on the protection of water against pollution caused by nitrates from agricultural sources was approved (now amended by Real Decreto 47/2022, *de 18 de enero, sobre protección de las aguas contra la contaminación difusa producida por los nitratos procedentes de fuentes agrarias*). The essence of this regulation was the transposition into Spanish domestic law of Council Directive 91/676/EEC concerning the protection of waters against pollution caused by nitrates from agricultural sources and to ensure compliance with its objectives.

This Royal Decree 47/2022 aims to improve the results obtained by the aforementioned regulations, which have not achieved their intended purpose. In order to strengthen Spain's commitments, along the lines indicated in the European Biodiversity Strategy 2030 and other related strategies, it is necessary to review the current forms of intervention in order to make them more efficient and effective. To this end, it is of interest to consider the experience acquired in recent years, the work and technical studies carried out, and the jurisprudence established by the Court of Justice of the EU. With all this in mind, the drafting of this regulation was undertaken, which replaces and broadens the ambition and scope of Royal Decree 261/1996 of 16 February 1996, the original regulation transposing Council Directive 91/676/EEC, which is repealed with this new regulatory instrument.

The Royal Law Decree 11/1995 of 28 December 1995, which establishes the rules applicable to urban wastewater treatment, should be treated separately. Its aim is to supplement the legal system established by the Water Law and to protect the quality of continental and marine waters against the adverse effects of urban wastewater discharge (Article 1). This article also defines key terms, such as urban wastewater (household wastewater or its mixture with industrial wastewater or with rainwater run-off), household wastewater (wastewater from residential and service establishment areas generated mainly by human metabolism and household activities), industrial wastewater (all wastewater removed from facilities used for any commercial or industrial activities, other than household wastewater and rainwater run-off), sensitive area (water area distinctly defined in accordance with the criteria to be defined by a royal decree), or a less sensitive area (part of marine waters or an area of marine waters delineated in accordance with the criteria to be defined by a royal decree).

The issue of integrated permits concerning wastewater discharge to waters is regulated by the Royal Law Decree 1/2016 of 16 December 2016, approving the consolidated text of the law of integrated pollution prevention and control.

Finally, in Spain, there are current guidelines of July 2020 (they are not standard and are not binding) that constitute the first state-level test to regulate groundwater pollution; all of this, without prejudice to the initiatives of some water authorities to also define, by means of technical guides, groundwater quality values. This is in addition to Law 7/2022 of 8 April on waste and contaminated soils for a circular economy, where it provides for the integration of the measures contained in this title in the programmes to be established for the protection of the marine environment and water and safeguards compliance with EU food legislation to ensure food hygiene and food safety (Companys Alet & Loste Madoz, 2020).

**Water authority – responsibilities and competences**

The water authority is divided between the state and autonomous communities. Notably, according to Article 2 of the Constitution of Spain, this act is based on the unbreakable unity of the Spanish nation, the common and indivisible homeland of all Spaniards, and it acknowledges and ensures the right to autonomy of the nationalities and regions that comprise it, and solidarity of all. Article 143, Section 1 of the Constitution of Spain states that – when exercising the right to autonomy, recognised in Article 2 of the Constitution – bordering provinces of common historical, cultural, and economic features, as well as insular territories and provinces that constitute regional historical unity, can join the local government and make autonomous communities, in accordance with the provisions of this title and the respective statutes. Provisions of Articles 148–150 define the borders and rules of the legislative activity of autonomous communities.

It must be understood that the rights of private use of water in perpetuity are not compatible, in terms of the non-purely formal effectiveness of the legal rules, with the principles of inalienability and imprescriptibly of the public domain property enshrined in Article 132.1 of the Constitution. Therefore, the temporary limitation of such private use is not a deprivation of rights but a new regulation of the same which does not affect their essential content. The freezing of the material substratum of the rights consolidated prior to the enactment of the Water Law in no way implies a partial expropriation of the same, as this only eliminates the simple expectations of exploitation of higher flows that could eventually be obtained due, on the one hand, to the ownership of the real property right, but also, and in necessary concurrence with this, to the character of 'res nullius' that the waters not flowing or illuminated had according to the previous legislation and to the non-existence or preference of third-party rights. In short, what the law excludes from now on is the possibility of the appropriation of property from possible increases in the flows used without a concessionary title (Judgement of Constitutional Court, STC 227/1988). Article 132.2, while directly excluding the private ownership of some types of property, allows the legislator to declare the demaniality of others. (In accordance with this constitutional provision, the option of including inland waters in the public domain is legitimate in any case).

However, since all renewable surface waters and groundwaters are transformed 'ex lege' into public domain waters, it is lawful that, while strictly respecting existing rights, increases in the appropriated flows can only be obtained by means of an administrative concession. The state has all the competences which, in accordance with the Constitution, have not been or have not been able to be assumed by the autonomous communities in their respective statutes, and as long as they do not assume them or they are transferred to them given that the statutory provisions on a future extension of the scope of autonomous competences do not constitute in themselves. So, the

Spanish Constitution sanctions a guarantee of the property, goods, and patrimonial rights of individuals (Article 33). But this guarantee is not absolute since Article 128.1 establishes that 'all the wealth of the country in its different forms is subordinated to the general interest', and, as far as is relevant here, Article 45.2 imposes on the public authorities the duty to ensure 'the rational use of all natural resources, with the aim of protecting and improving the quality of life and defending and restoring the environment, based on the indispensable collective solidarity'.

According to Article 1, the aim of the Water Law is to provide legal regulations within the state competence specified in Article 149 of the Constitution of Spain of 27 December 1978. Section 1 contains provisions concerning issues being within the exclusive competence of the state, and Section 3 contains a norm of competence and states that issues not specifically reserved for the state by the Constitution can be passed on to the competence of autonomous communities under their statutes. The competence in matters which would not be taken over under statutes on autonomy lies with the state, whose law – in cases of collision – takes precedence over the laws of autonomous communities in everything which is not their exclusive competence, and the state law supports and complements the law of autonomous communities (Judgement Spanish Constitutional Court. STC 227/1988 (La Ley 2428/1988)), it is stated that the state has competence only with regard to intercommunity basins to draw up the corresponding Hydrological Plan, Articles 100 and 101 of Regulation 927/1988. For example, the Statute of Andalusia adopted by the organic law 2/2007 of 19 March 2007, in its Article 50, grants to this autonomous community the competence concerning some cases related to waters.

The following regional norms, contained in the amended Act 7/2007 of Andalusia of 9 July 2007 on integrated environmental quality management apply (Ley 7/2007, Ley 1/2008, Ley 4/2010, Decreto 356/2010, Ley 9/2010, Decreto 239/2011) and the act on water 9/2010 in Andalusia of 30 July 2010, are applicable in the field of water management and protection. Because of the drought of recent years, a special Law Decree 3/2023 of 25 April is in force (Marcos Fernández, 2010).

However, when establishing the applicable water law, the Spanish legislators made an *a priori* assumption on the functioning of various autonomous communities, dividing the water authority into the state functions and tasks of other entities.

In terms of the state functions, the Water Law contains various references to the general state administration (*Administración General del Estado*) or directly to the minister (*Ministro de Medio Ambiente*) or the ministry (*Ministerio de Medio Ambiente, Ministerio de Agricultura, Alimentación y Medio Ambiente*).

Organisation of the general state administration is defined in the Act 40/2015 of 1 October 2015 on the public sector legal system. It identifies central administration, which is divided into authorities (ministers and secretaries of state) and governing organs (undersecretaries of state and general secretaries, provisional general secretaries, general directors, and deputy general directors). There is also territorial administration, but only with governing organs: government delegates in autonomous communities and government subdelegates in provinces.

Minister of Ecological Transformations and Demographic Challenges (*Ministerio para la Transición Ecológica y el Reto Demográfico*) should be regarded as the supreme state water authority.

Functions of the National Water Management Board, which is the supreme consulting and participative body in this regard, are also provided for. The National Water Management Board comprises representatives of the general state administration, autonomous communities, all local governments through the agency of a representative of the most common association at the state level, organs of a drainage basin organisation, representatives of the most representative professional and economic organisations at the state level related to various uses of water, representatives of the most representative trade unions and business organisations at the state level, representatives of the most representative non-profit organisations at the state level, whose aim is to protect the environmental interests. Its composition and the organisational structure are set out in the royal decree, and the National Water Management Board is chaired by the minister under the relevant law.

Drainage basin organisations, whose functions and responsibilities are regulated by the Water Law, are created in functionally described drainage basins, exceeding the territorial range of an autonomous community. According to Article 22, drainage basin organisations are autonomous with respect to managing and administering the matters entrusted to them in their own name. And they have jurisdiction over administrative proceedings. The territorial range of a drainage basin organisation covers one or more indivisible, functionally understood drainage basins, with limitations arising from international borders. The local jurisdiction of a specific unit of this type is set out by the royal decree.

Drainage basin organisations have the following functions (a) developing a water management plan for a drainage basin, as well as monitoring and reviewing it; (b) administering and controlling public water areas; (c) administration and control of water use in the common interest or ones which have an impact on more than one autonomous community; (d) designing, construction, and exploitation of work done with the organ's own resources and those entrusted by the state; and (e) arising from agreements concluded with autonomous communities, local communities and other public or private entities or agreements concluded with individuals.

In order to perform the functions from the last two items, drainage basins can acquire, by subscription or purchase; divest; and perform any administrative activities with respect to stock representing capital- of state-owned companies, which were created in order to construct, exploit, or carry out public water work, or commercial companies, whose corporate goal is to manage concession agreements for construction and exploitation of water structures on condition of obtaining prior consent from the Ministry of Finance and concluding cooperation agreements or membership of groups of companies and participation in temporary joint enterprises whose objective is any one of the aforementioned, as well as granting loans and credits to the entities mentioned earlier (Article 23).

In order to perform their functions, drainage basin organisations have the following rights and obligations: (a) to grant permits and concessions related to public waters, except those related to work and actions in the state's general interest, which will satisfy the Ministry of Ecological Transformations and Demographic Challenges; (b) to control and monitor compliance with conditions of concessions and permits concerning public waters; (c) to conduct measurements, hydrological tests, information on floods and water quality control; (d) to test, design, execute, maintain, exploit and improve work included in their own plans, and any other that may be entrusted to them; (e) to determine goals and quality programmes in accordance with water planning; (f) to implement, within their competence, plans, programmes, and actions aimed at proper demand management in order to promote saving and economic and environmental efficiency in various water uses through global and integrated use of surface and ground water, in accordance with the provisions of the relevant sectoral planning, as applicable; and (g) providing all kinds of technical services associated with pursuing specific goals and, on request, giving advice to the general state administration, autonomous communities, local communities, and other public or private entities, as well as to individuals.

Drainage basin organisations are affiliated with hydrographic confederacies, which are autonomous organisations. Pursuant to Article 25 of the Water Law, drainage basin organisations can establish cooperation in exercising their competences, particularly by including representatives of autonomous communities in the Governing Boards of these organisations. Drainage basin organisations can conclude cooperation agreements with autonomous communities, local communities, and water cooperatives to exercise their respective competences as per the applicable laws.

If acts or plans of autonomous communities or local governments include new demand for water resources, a hydrographic confederacy report clearly specifies whether sufficient resources exist to satisfy such demand.

Pursuant to Article 26, the Governing Board and the chairperson are the governing bodies of drainage basin organisations. On the participation principle, the following are the governing bodies for exercising the functions clearly assigned to them by the law: Users' Assemblies, Discharge Committees, Exploitation Boards, and Site Boards. In turn, the Demarcation Water Board is a participation and planning body. On the other hand, the Committee of Relevant Authorities is a cooperation body with respect to the obligations arising from the Water Law in terms of water protection.

The Governing Board is responsible for (a) approving action plans of organisations, making budget proposals, and informing about their liquidation; (b) agreeing, as applicable, the necessary credit operations for specific objectives associated with management, as well as financing actions included in action plans within the boundaries specified by law; (c) passing resolution concerning the disposal of the property of drainage basin organisation; (d) preparing matters submitted to the Drainage Basin Board; (e) approving, based on a report of the Demarcation Water Board, changes of the width of easement belts and the protection zones; (f) notifying ground water bodies threatened with failing to achieve a good quantitative or chemical status and measures aimed at correcting the tendencies threatening a good status by approving the action plan for improvement without detriment to those which satisfy other public administration organs; (g) accepting decisions concerning water cooperatives; (h) promoting initiatives concerning wetlands; (i) proposing, at the chairperson's initiative, penalties for serious or very serious violations, when the facts are of particular importance for proper management of resources in a drainage basin; (j) approving, as applicable, the general criteria of establishing compensation for damage to public waters; (k) putting forward proposals to the Demarcation Water Board to revise the relevant water plan; (l) considering issues submitted for consideration by the chairperson of the board or any of its members (Article 28).

It is the task of the Exploitation Board to coordinate while respecting the rights arising from the relevant concessions and permits, the exploitation of hydrotechnical devices and water resources of a group of rivers, a river, a river section, or a hydrogeological unit whose methods of use are particularly linked to each other. Proposals of the Exploitation Boards with respect to their competences are passed on to the chairperson of the drainage basin organisation. The Exploitation Board is composed of the users (Article 32).

According to Article 33, the Discharge Committee is responsible for the meeting agenda and for putting forwards motions to the chairperson of the body concerning the relevant system of filling and emptying the water bodies and aquifers, considering the concession rights of various users.

The Board of Directors, upon request of future users of the work already approved, can appoint a relevant Site Board, of which such users can be members so that they are directly informed about the developments associated with this work (Article 34).

A Water Drainage Basin Board is appointed to inform, promote, and carry out social consultation and active participation in the planning of water management in the drainage river areas with intercommunal drainage basins. Its obligation is to submit a water plan for the drainage basin and its subsequent modifications to the government through the ministry. Representatives of autonomous communities whose territories are partly or wholly a part of the drainage basin are included in the respective Water Boards so that they participate in preparing the water plan and executing other functions of the board. The composition is established by the royal decree in accordance with the following criteria: (a) each ministerial department associated with the water management and using water resources is represented by up to three members; (b) technical services of the drainage basin organisation will be represented by up to three members; each peripheral coastal service of the ministry, whose territory coincides partly or wholly with the drainage basin will be represented by one member; each harbour master's office concerned with the drainage basin will be represented by one member; (c) representation of autonomous communities – members of the board – is established and distributed according to their number in the drainage basin and their area and the population size, with each of the participating autonomous community being represented by at least one member, and the total representation of the autonomous communities cannot be lower than that corresponding to various ministerial departments; (d) local authorities, whose territory coincides fully or partly with the drainage basin will be represented according to the expansion or percentage of the territory, which is affected by the drainage basin, on conditions to be specified in the regulation. The maximum number of members cannot exceed three; (e) a representation of users cannot be smaller than a third of the total number of the members, and it comprises representatives of various sectors with respect to their respective interests in using waters; and (f) representation of the associations and organisations acting in protection of environmental, economic and social interests related to water. The maximum number of members is six. For drainage basins in intra-community drainage basins, the respective autonomous community guarantees community participation in water planning while observing these minimum representations of users and organisations concerned in collegial organs created for this purpose and ensuring that all public administration bodies with authorisations in matters related to water protection and, in particular, with respect to the general state administration authorisations related to public land marine areas, public utility harbours, and merchant ships are equally represented in these bodies (Articles 35–36).

Committees of Relevant Authorities of Drainage Basin Areas operate in order to ensure proper cooperation in implementing regulations on water protection for drainage basin areas covering intercommunal drainage basins. Their creation does not affect the exercising of competence corresponding to various public administration bodies in matters related to water management, which will still be performed in accordance with the applicable regulations in each case. The following are the basic functions of Committees of Relevant Authorities of Drainage Basin Area: (a) promoting cooperation in exercising competences related to water protection held by various public administration bodies in a specific drainage basin area, (b) promoting the adoption of measures required for compliance with the protective provisions of this act by the public administration bodies competent in each area, and (c) providing information on the drainage basin area, which may be required under applicable regulations, to the EU through the Ministry of Environment. The committee comprises (a) general state administration bodies with competence for use, protection, and control of waters with the number of representatives not exceeding the number of representatives of autonomous communities; (b) organs of autonomous communities whose territory is included in the drainage basin area in whole or in part, with authorisations in protection and control of waters covered by this act, one representative of each autonomous community; and (c) local authorities whose territory coincides in whole or in part with the drainage basin area, which has authorisations concerning protection and control of waters, which is represented in accordance with the size of the population in the area, through the respective territorial commune associations. For intra-community drainage basin areas, the respective autonomous communities guarantee the uniformity of water management cooperation in exercising the competences held by various public administration bodies with respect to their protection, in particular of those corresponding to general state administration in matters related to marine, harbour, and trade public domains. They also provide to the EU, through the ministry, information concerning the drainage basin area in accordance with the applicable regulations (Article 36 bis).

## Water police as a control organ in surface water management

Article 94 of the Water Law applies to water police. In this regard, supervision over waters and other elements of public waters, easement areas, and protected areas is exercised by the competent water administration body. Water Committees in drainage basin areas exceeding the territorial range of an autonomous community perform the following functions: (a) inspect and control public water areas; (b) inspect and monitor compliance with the conditions of concessions and permits related to public waters; (c) perform measurements, inform about floods, and control water quality; (d) inspect and monitor work arising from concessions and permits related to public water areas; (e) control and supervise exploitation of all public water structures, regardless of their ownership and legal regime to which they are subjected; (f) manage river guard; and (g) apply regulations concerning water and watercourse protection. Environmental protection personnel assigned to river police stations in river basin organs hold the status of public authorities when performing their functions, and they are authorised to (a) enter freely and at any time, without giving advance notice, places to be inspected and to stay there, while respecting the inviolability of domestic peace. While conducting an inspection, the personnel inform the person being controlled or their representative about their presence unless they believe that such information may undermine the performance of their duty. (b) Perform any investigations, examinations, and tests which they will regard as necessary to check whether legal regulations are followed correctly. (c) Collect samples of substances and materials used or owned by a site, perform measurements, take photographs, record video clips, record images, and make sketches and drawings, provided the employer of his representative is informed. The facts established by environmental protection personnel, provided in the relevant reports, are regarded as accurate, without prejudice to evidence, which the parties concerned can submit in defence of their rights and interests. The river guard performs tasks that involve supporting and providing help to environmental protection personnel in performing their functions related to water protection.

The system of sanctions provided for in the law also regulates the issues related to public water protection and water quality. According to Article 116 of the Water Law, actions that constitute a violation are understood broadly – as failure to comply with any of its provisions. They are punishable by administrative penalties under Act 39/2015 on joint administrative proceedings in public administration. Interestingly, the Water Law regulates joint responsibility, and it is applied when several individuals are responsible, and it is impossible to determine each one's contribution to the violation. The following are regarded as administrative violations: (a) actions that result in damage in the public water domain and hydrotechnical structures; (b) water intake from watercourses and groundwater intake without a proper concession or permit, if one is required; (c) failure to meet the conditions laid down in concessions and administrative permits, without prejudice to their expiry, rescinding or suspension; (d) executing without a proper administrative permit, other work, sowing or planting in public water courses on in areas with limitations with respect to their intended use or use; (e) interference, acquisition, or excavation of aggregate from water courses without a proper permit; and (f) performed without a proper permit, water discharges which can lower the quality of water or the conditions of the drainage of the water course to which it is discharged.

## Summary

Spain is a regional country, which has its consequences in the specificity of the sources of the law concerning water management. There are state laws implementing the WFD, and there are specific regulations passed by autonomous communities. In the national legal framework, the subject matter concerning water law is basically concentrated in one normative act and in executive regulations. The hydrological situation of Spain necessitated centralisation and, at the same time, deconcentration of the functional and institutional system. Planned and targeted water management has been conducted in this country for many years, and water plans are of a normative nature. Many public entities are involved in the process of adopting them, and the planning procedures contain participative and consultative mechanisms. Water plans are coordination measures and provide the basis for rational water management and water resource protection. Apart from public law instruments, there are traditional private law institutions, which make them comprehensive. They also constitute the basis for issuing permits and concessions and provide a point of reference for courts and tribunals. The institutional framework for water management is not only established by the functions of the state, including the adopting the national Water Law and tasks of autonomous communities, and the scope of activities of provinces (*provincias*) and municipalities (*municipios*) in matters related to water management is limited to tasks performed by cooperation. On the one hand, specialisation brought about the establishment of extremely professional and structurally complex drainage basin organisations, which carry the burden of water resources management and surface water protection. The entities have various administrative competences, but they also exercise their authority both in the sphere of *dominium* and fiscal measures. On the other hand, traditional rules of joint use of such resources by individual users (farmers) and the operation of irrigation cooperatives, which are sometimes as old as 1,200 years, are still valid in Spain. Irrigation cooperatives can operate under state and regional regulations, but – as the case of Valencia shows – only on the basis of common law. The legal system of surface water management and protection in Spain should be regarded in a multi-centre manner, and it comprises many elements of a diverse normative nature. The legal solutions concerning the Water Law are extremely stable. However, the drought of recent years has brought extraordinary legal solutions, which – through consecutive renewals – are becoming permanent. However, nowadays, groundwater has been exploited uncontrollably, which has resulted in more water being taken illegally than from legal sources, which results in a crisis of the existing system solutions. Community control through irrigation cooperatives and resolving disputes as part of the joint use of common resources (*common-pool resources*) has become ineffective. *De lege ferenda* proposes that the operation of drainage basin organisations should be nationalised further and its scope expanded to include control and IT measures.

## References

Agudo González, J. (2010). El derecho de aguas en perspectiva europea: la trascendencia de la Directiva Marco de Aguas. In *El derecho de aguas en clave europea*, edited by J. Agudo González, 69–141. La ley.

Alcaín Martínez, E. (2010). La protección jurídico-privada del derecho de aprovechamiento de aguas. *Diario La Ley*, 7366.

Bunclark, L., R. Carter, V. Casey, S. J. Day, and D. Guthrie (2011). *Managing water locally: An essential dimension of community water development*. Oxfam UK/WaterAid/Institute of Civil Engineers.

Chiu, V. (2014). *La protection de l'eau en droit public: Étude comparée des droits espagnol, français et italien*. Droit: Université de Toulon.

Companys Alet, A., and J.A. Loste Madoz (2020). Contribución al nuevo régimen de suelos y aguas subterráneas contaminadas. Problemática derivada de la aplicación conjunta del régimen de suelos contaminados y de residuos. *Diario La Ley*, 9699.

Döpp, L. (2000). *Das spanische Wasserrecht: die Verteilung von Wasser als Regelungsproblem*. Münsteraner Studien zur Rechtsvergleichung, 74.

Embid Irujo, A. (2017). Organizaciones de usuarios y participación en la gestión del agua en el Derecho español. In *Perfiles de la ordenación jurídica del agua en Italia*, edited by G. Crepaldi, 45–62. Valencia: España y América Latina.

Embid Irujo, A. (2009). Water Law in Spain After 1985. *Water International* 28 (3):290–294. https://doi.org/10.1080/02508060308691701

Fanlo Loras, A. (2006). Le modèle espagnol de participation du public à la gestion de l'eau: mythe, réalité et défis immédiats. *Environnement* 7:51–63.

Flores García, A.I., and M.J. Molina García (2017). Tratamiento legal de las aguas subterráneas en España: ¿Se garantiza un uso sostenible de este recurso natural? Diario La Ley, 8971.

Gallego Anabitarte, A., A. Menéndez Rexach, and J.M. Díaz Lema (1986). *El derecho de aguas en España*. Ministerio de Obras Públicas y Urbanismo, Secretaría General Técnica.

González Pérez, J., J. Toledo Jáudenes, and C. Arrieta Álvarez (1987). *Comentarios a la Ley de aguas*. Civitas.

Guaita Martorell, A. (1982). *Derecho administrativoaguas, montes, minas*. Civitas.

Hernández, L.S.M., and J.C. González Martínez (2023). Reutilización de aguas regeneradas en la cuenca del Segura. Adaptación al reglamento (UE) 2020/74: retos y oportunidades. In V Congreso Nacional del Agua - Libro de Actas, edited by J. Melgarejo, M. López Ortiz, and P. Fernández Aracil. Alicante, 2023.

Maestre Rosa, J. (1969). *Comunidades de regantes. Concepto, naturaleza jurídica y regulación positiva*. Bosch.

Marcos Fernández, A. (2010). Las demarcaciones hidrográficas en la Directiva Marco de Aguas y su desarrollo en el Derecho español. In *El Derecho de aguas en clave europea*, edited by J. Agudo González, 186–250. La ley.

Martín-Retortillo Baquer, S. (1960). La elaboración de la Ley de aguas de 1866. *Revista de Administración Pública* 32:11–54.

Menéndez Rexach, Á. (2010). El derecho al agua en la legislación Española. In *El derecho de aguas en clave europea*, edited by J. Agudo González, 53–84. AFDUC.

Molina Giménez, A. (2021). Análisis jurídico del Reglamento (UE) 2020/741, de 25 de mayo de 2020, sobre reutilización de aguas regeneradas, y estudio de su repercusión en España. *Revista Aranzadi de derecho ambiental* 48:147–202.

Molina Giménez, A. (2022). La evaluación y gestión del riesgo que conlleva la actividad de reutilización. *Observatorio del ciclo del agua* 2:311–343.

Organización de las Naciones Unidas para la Agricultura y la Alimentación (1976). *El derecho de aguas en algunos países europeos*, Vol. 1. Bélgica, España, Francia, Inglaterra y País de Gales, Israel, Italia, Turquía: Roma.

Sánchez Camacho, R. (2007). Las competencias sobre el agua en las reformas estatutarias. *Aletheia: Cuadernos Críticos Del Derecho* 2:58–78.

Schwarz, E. (1966). *Das spanische Wasserrecht: geschichtliche Gundzüge, Eigentumsverhältnisse an den Gewässern, Nutzungsverhältnisse, Dienstbarkeiten*. Fotodruck Präzis.

Vergara Blanco, A. (1991). Contribución a la historia del Derecho de Aguas, II: Fuentes y principios del Derecho de Aguas español medieval y moderno. *Revista de minas y aguas* 2:137–161.

Wawer, R. (2021). Agricultural water management in changing climate – a call for a reform in Poland. Case analysis of reference – a 1200 years of irrigation communities in Spain. In *13th International Conference on Agrophysics: Agriculture in changing climate. Lublin.*

# 5 Legal protection of surface waters in Lithuania

*Jakub Jan Zięty and Kristina Mikalauskaitė-Šostakienė*

## Introduction

The analysis presented in this part of the study deals with legal regulations applicable to surface water management and protection in the Republic of Lithuania. Provisions of the Water Framework Directive (WFD) were implemented into the Lithuanian Water Law of 1997. The main regulations which may concern environmental protection, with water protection and management, include the following:

1. The 1997 Water Law of the Republic of Lithuania of 1997 (*Lietuvos Respublikos Vandens Įstatymas*)
2. The Environmental Protection Law of the Republic of Lithuania of 1992 (*Lietuvos Respublikos Aplinkos Apsaugos Įstatymas*)
3. The Marine Environmental Protection Law of the Republic of Lithuania of 1997 (*Lietuvos Respublikos Jūros Aplinkos Apsaugos Įstatymas*)
4. The Law of the Republic of Lithuania on environmental impact assessment for planned business activities of 1996 (*Lietuvos Respublikos Planuojamos Ūkinės Veiklos Poveikio Aplinkai Vertinimo Įstatymas*)
5. The Law of environmental monitoring in the Republic of Lithuania of 1997 (*Lietuvos Respublikos Aplinkos Monitoringo Įstatymas*)
6. The Law of the Republic of Lithuania concerning waste management of 1998 (Lietuvos Respublikos Atliekų Tvarkymo Įstatymas)

The legal protection of surface waters is carried out in drainage basins. There are four main drainage basin districts identified in the Republic of Lithuania (Nemunas – LT1100, Venta – LT2300, Lielupe – LT3400, and Dauguva – LT4500). These drainage basins were identified to manage waters in the drainage basins and to implement the WFD (Zębek, 2018). Implementation of measures provided for in the WFD is planned in all the River Basin Districts (RBD). However, these laws were implemented on the basis of domestic laws by virtue of a domestic regulation, i.e., the development programme for the water sector for the years 2017–2023 and the action plan for implementation of the water sector development for the years 2017–2023. This programme

DOI: 10.4324/9781003586913-6

sets the objectives of the water sector and the action plan for the implementation of the water sector development programme, including the measures for attaining this goal, sources of financing, and the measure implementation mechanisms. Moreover, the agricultural measures 2014–2020 were implemented in order to improve the quality and to protect waters as part of the rural areas development programme.

Currently, a third management plan for the river basins of the Neman, the Lielupe, the Venta, and the Daugava for the years 2022–2027 has been prepared and approved in order to improve the water quality, and this was the base for preparing and approving the National Water Plan for 2022–2027 (2022, Nr 1292) and its executive action plan. A further part involves taking actions aimed at attaining the goals set in the WFD. When developing plans for the third period of the UBR (UPIŲ BASEINŲ RAJONO – RIVER BASIN DISTRICT) for the Neman, the Lielupe, the Venta and the Daugava, 1,193 surface water bodies were identified, of which 826 were classified as rivers, 361 – as lakes, 4 – as intermediate (Curonian Lagoon, Curonian lagoons spread over the Baltic Sea) and two coastal ones (territorial waters within one mile of the coast) categories of water bodies, with 38 water reservoirs are isolated. The assessment criteria for the ecological status of water bodies, lakes, coastal waters, and artificial and highly modified water bodies are set out in the methodology for determination of surface water body status, approved by the Minister of the Environment. The aim of these plans is to improve the status of surface water bodies and groundwater bodies, and particular attention will be paid to controlling dispersed pollution and 'liberating' the rivers. Its aim is also to reduce the amounts of pollutants reaching the Baltic Sea and the Curonian Lagoon with river run-off, to improve the preparedness for floods and prevent them. Moreover, there are plans to improve the effectiveness of the water body management system. The plan also covers another important area – high-quality assurance with respect to drinking water supply and wastewater management, as well as the reduction of environmental pollution by wastewater. Plans to this end include improving the efficiency of water supply to enterprises and implementing the cost reimbursement rule.

## Legal basis of surface water management and protection

Provisions of the Lithuanian Water Law regulate the ownership of inland water bodies in the Republic of Lithuania, the management, use, and protection of water resources, relationships between the owners and users of water bodies, the rights, and obligations of legal entities and individuals using water resources and water authorities (regulatory bodies). Provisions of this law apply to all inland water bodies and natural resources in the Republic of Lithuania, regardless of their intended use and forms of ownership. The law defines the minimum water flow throughput, which guarantees ecosystem maintenance, as well as such terms as drinking and mineral water, the hydrographic system, water structures, wastewater, wastewater treatment plants, pollution level, water quality, water quality standards, water owners, water supply, and watercourses (Pietkiewicz et al., 2020).

Water resources constitute an integral part of the Republic of Lithuania's natural resources. They have to be used in accordance with the procedure laid down in the Constitution of the Republic of Lithuania, the Environmental Protection Act, and other laws. Importantly, water resources, as a collective term, covers waters owned by the National Inland Water Fund and private entities. Waters can be owned only by state entities and citizens of the Republic of Lithuania, while groundwaters are owned exclusively by the state. In the case of surface waters, those owned by the State Inland Navigation Fund include (1) Curonian Lagoon; (2) rivers with constant outflow, with a river basin area of at least 25 m,[2] 1 km of which connect lakes assigned to the State Inland Navigation Fund; (3) lakes connected by rivers transferred to the State Inland Navigation Fund, as well as lakes transferred to the State Inland Navigation Fund by the government or a body authorised by it; (4) potentially dangerous ponds with an area in excess of 5 ha or a dam height in excess of 3 m; and (5) all the other non-private water bodies. A list of inland waters owned by the state, even if they do not meet these criteria and are inland waters of national importance, is drawn up and approved by the government of Lithuania.

Water resource users include legal entities, which (1) leased surface water bodies for a specific use area on conditions set in agreements in accordance with procedures set out by the government or by an authority authorised by it, (2) obtain water from water bodies or discharge wastewater to the natural environment, and (3) use water reservoirs and resources for hydro energy, navigation, as fish ponds.

Users of water resources (individuals and legal entities which use water resources) have a right to (1) use water resources to conduct activities regulated by the law, (2) dispose of the revenue from water resource use, and (3) in cases and on conditions specified by the law or in agreement with the authorities authorised by the government, allow using water resources by other legal persons and individuals under an agreement.

Water resources and reservoirs can be used to supply drinking water to the population, for recreation, sporting, agricultural, industrial, and other business activities, navigation, hydropower, fisheries, hunting, extinguishing fires, wastewater removal, and for other purposes provided it is not forbidden by the laws of the Republic of Lithuania. Supplying drinking water to the populations is the primary use of water resources in water bodies. The Lithuanian Water Law provides for common (angling, swimming) and free (for their owners) and paid use of water. Importantly, monitoring water quality is an element of environmental monitoring, and it is described in the Act on Environmental Monitoring.

The Lithuanian law provides for several types of legal instruments concerning surface water management and protection. These instruments include the monitoring systems mentioned earlier and the system of issuing environment use permits, including permits for water use for various purposes, including business activities or control of water use. Apart from those mechanisms, surface water protection was established through the following instruments: (1) planning – state-wide and regional plans and water management programmes, (2) administrative – setting out the general obligations and bans for water users and a system of permits and binding opinions (permit, consent, and opinion), (3) informative – a monitoring system for the quantity and quality of waters, (4) corrective – remedial and corrective in cases when water is polluted as a result of a failure, and (5) special – in the event of a flood.

Users of water bodies assigned to the State Inland Navigation Fund and owners of private water bodies have to (1) meet the requirements concerning the maintenance, repair, and reconstruction of ponds and their hydrotechnical structure in order to ensure the stability of surface water body banks. Maintenance and use of potentially dangerous ponds and their hydrotechnical structures are determined in the act on the maintenance of potentially dangerous objects and other acts. (2) Pursuant to the regulatory requirements, establish a strict regime for the sanitary protection zone of a watershed and maintain it. (3) Regulations require that coastal protection zones should be established for surface water bodies in accordance with legal regulations and that such belts and protective zones should be maintained. (4) Do not change the water level in rivers or lakes without the consent of a body authorised by the Ministry of Environment Protection, not carry out river dredging or regulation, not clean the bottoms or modify the banks, not build dams or other hydrotechnical structures. (5) Use water effectively; do not exceed the water use limits and norms of wastewater and pollutants discharged into the environment. (6) Do not infringe upon other water resource users' rights and not act to the detriment of the landscape or its elements (land relief, soil, flora, fauna), subsoil, agricultural, and agritourist farms or other objects. (7) Do not destroy aquatic flora or fauna. (8) Maintain records of water consumption in a specific manner and conduct monitoring in accordance with the Act on Environmental Monitoring. (9) Do not conduct any activities on inland water bodies/routes designated as marine navigation routes which would jeopardise navigation or constitute water routes.

Water resources can be used in return for a fee or free of charge. If water is used in return for a fee, conditions for such use are specified in a permit. Permits for water use for legal entities and individuals are issued in accordance with a procedure established by the government or an institution authorised by it. Permits for the use of water resources are not granted unless resources are allocated that allow for rational water use, effective protection of water bodies against pollution, proper conditions for aquatic organisms, and the prevention of the infringement of the interests of other water users.

Water monitoring is one of such instruments. An assessment conducted in accordance with the methodology of determining the status of surface water bodies is based on an assessment of a six-year period of the Management Plan for Waters of a River Basin District, and it is comprehensive. It covers not only the physicochemical quality parameters (nutrients, organic substances, saturation with oxygen, clarity of water, specific pollutants) and biological quality parameters (aquatic flora, phytoplankton, invertebrates, fish) but also hydromorphological quality elements (hydrological regime, river integrity parameters, morphological conditions) and hazardous substances. It is not the status of surface water bodies that can be assessed annually but water quality according to individual quality parameters. The methodology of determination of the status of surface water bodies approved by ordinance No. D2007-12 is applied in the assessment of the status of surface water bodies for which goals for protection of waters were established, i.e., lakes (including those classified as heavily modified water bodies, ponds, stone pits) with an area exceeding 0.5 $km^2$ and rivers (including those classified as heavily modified water bodies and canals with the drainage basin area exceeding 30 $km^2$).

According to the State Environment Monitoring Programme, state monitoring of lakes and rivers is conducted annually, which includes measuring physicochemical quality parameters, biological quality parameters (phytoplankton, aquatic flora, macrophytes, phytobenthos), invertebrates (macrozoobenthos), fish) and hazardous substances. Mean annual data for physicochemical quality parameters and hazardous substances and data concerning water quality assessment through biological parameters are published on the Environmental Protection Agency's (EPA) website (2024).

Specific financial instruments are provided for in the Lithuanian law for water use. Legal entities and individuals pay fees for using water resources and their pollution in accordance with the acts on payments for state natural resources, taxes on environmental pollution, and other laws. The government can, in accordance with the procedure established by the law, compensate legal entities and individuals for the cost of water supply, as well as the construction, use and maintenance of wastewater treatment plants, if these facilities are built for state or public funds.

The Lithuanian Water Law allows the use of water bodies for surface (rainfall) wastewater removal. This also applies to other wastewater generated in households as a result of industrial and other business activities, sewers, and other wastewater (Plunge et al., 2022; Plunge et al., 2023). However, this is possible upon the consent (permit) of an organ authorised by the Ministry of Environment Protection and within the limits of the water quality standards of the wastewater being removed, as well as the water in the water bodies, which are approved by the Ministry of Environment Protection in agreement with the Ministry of Health.

According to Lithuanian law, an environmental impact assessment for an enterprise provides for community participation in such procedures. One or more individuals or legal entities, a community concerned, have a right to (1) obtain information on the environment in accordance with the established procedures; (2) participate in the procedures of an environmental impact assessment and an environmental impact assessment for planned business activities in accordance with the established procedures and file petitions; (3) demand discontinuation of harmful environmental impact of a business activity; (4) organise and participate in public environment control; (5) require the authorities and state administration to organise environmental education and training and to propagate freely environmental protection ideas; 6) file a complaint (petition) in accordance with the procedure established by the law of the Republic of Lithuania and demand that proper measures be taken in order to prevent or control the damage to the environment or to restore the original status of the environment and to demand that individuals guilty of a harmful impact on the environment, officials whose decisions or actions (inaction) infringed on rights of citizens, community concerns, other legal entities, and individuals or interests protected by law should be punished; and (7) pursuant to the procedure established by the law of the Republic of Lithuania, to petition a court, if they find that their petition, filed in accordance with the procedure established by the laws regulating the right to obtain information on the environment was unlawfully rejected, was considered incorrectly in whole or in part or was not properly granted in accordance with the laws regulating the right to obtain information on the environment. A person concerned has the right to seek court protection of the public interest in accordance with the procedure established in the laws of the Republic of Lithuania and to question the legality of the decision, actions, or inaction concerning the environment, its protection, and use of natural resources on material or procedural grounds.

As part of various proceedings for consent to the use of environmental resources, organs are obliged to make the environmental impact assessment public. The public has the right to obtain information from the entities conducting an environmental impact assessment for planned business activities on the method applied in the environmental impact assessment (including for surface waters), to file petitions, comments, and reservations, which the entity planning to conduct business activities has to consider and justify its decision. Community participation in the process of the initial environmental impact assessment for planned business activities is organised by communes. A public hearing concerning the results of a detailed environmental impact assessment for planned business activities is organised by the developer at his own expense.

Using water resources contrary to the regulations is sanctioned. Individuals are subject to disciplinary, administrative or penal responsibility under the laws of the Republic of Lithuania if they (1) exploit water resources without a permit when one is required; (2) violate the regime for hydrographic, hydrological, ichthyological reserves; (3) approve project documentation which does not provide for measures aimed at reducing any adverse impact on water bodies and resources if such should have been foreseen; (4) fail to implement measures aimed at water protection provided for in state programmes, plans or permits for water resource use; (5) for construction, modification, putting to use and accepting for use of business or other activity establishments, which can have an impact on water resources and water bodies in violation of a procedure established by the law; (6) pollute water reservoirs or their basins with wastewater exceeding the limits for pollutants, as well as the standards set in regulatory documents, cause thermal pollution of waters; (7) pollute water bodies or their basins in excess of the limits by household waste, production waste, or other waste, petrochemical products, fertilisers, toxic chemicals, radioactive substances, or other substances harmful to humans; (8) violate the regime of water protection in water bodies or water courses; (9) violate the regime of protective belts or zones for water bodies or water courses, pollute the banks of surface water bodies; (10) violate the rules for accounting for water use, determination of the quality of removed wastewater, wastewater treatment plants, use and maintenance of hydrotechnical structures, ponds; (11) damage installations or equipment in hydrotechnical structures, do hydrotechnical work by themselves; (12) violate established procedures for providing information on use and protection of water resources or conceal, fail to provide, or distort information; (13) interfere with state control of water resource use and protection, supervision of hydrotechnical structures, disregard lawful instructions given by controlling officials; and (14) commit other violations of the regulations concerning water use or its protection, for which they can be held liable in accordance with the law of the Republic of Lithuania. Apart from these sanctions, Lithuanian law provides for liability for damages for the harm done as a result of an offence (Klimaszewska et al., 2017).

## Water authority – responsibilities and competences

Water resource management is conducted by the government or the organs authorised by it in accordance with the rules laid down in the Lithuanian Water Law and other laws of the Republic of Lithuania. The use of water resources of the Republic of Lithuania on behalf of the state is regulated by the Parliament, Ministry of Environmental Protection, and other communal organs and institutions in accordance with their competences set out in the laws and regulations of the Republic of Lithuania. The state control over water use and protection is exercised by the Ministry of Environmental Protection and the EPA. Other state institutions control water use and protection within the scope of functions entrusted to them by the law of the Republic of Lithuania and government resolutions. The supervisory and control tasks, as well as issuing permits within the Ministry of the Environment, are the responsibility of the Department of Environmental Protection. In terms of the environment, it exercises the state control over environmental protection: it checks whether individuals and legal entities meet the requirements laid down in the Environmental Protection Act and other environmental regulations concerning the prevention of pollution; managing waste (except radioactive waste); accounting for, and sanitary protection of, wood raw materials; performing the obligations of producers/importers, entities placing wood and wood products on the market; using chemicals and purposeful release to the environment of organisms genetically modified outside the environment with the intention of placing them on the market; controlling limited use of genetically modified organisms; protecting landscapes, forests, biodiversity; carrying out control of dangerous objects wherever provided for by the law; and monitoring the correctness of calculation and declaration of environmental protection taxes. It considers, within its competence, petitions, complaints or reports from individuals and conducts preventive actions aimed at ensuring that individuals and legal entities observe statutory and executive requirements that regulate environmental protection and natural resource use, including surface waters (Dvarioniene & Stasiskiene, 2007).

The EPA is an organ for environmental control and monitoring. The agency is a state legal entity financed by the state treasury. The agency is managed by a director, and it is divided into departments that deal with various areas of the agency's activities (Plunge & Piniewski, 2024).

The agency performs tasks in the area of monitoring as well as environmental control, including surface water quality. The agency's mission within its competence includes the implementation of measures aimed at maintaining and improving the status of surface water bodies; accumulation, assessment, and/or other management; provision of information about the status of the environment, emissions of substances, and preparations; implementation of pollution preventive measures, including for those flowing into surface waters; implementation of environmental regulations in business activities; and implementation of chemicals and preparation (mixture) management and ensuring the rational use of live natural resources and their proper protection which meets the legal requirements. Moreover, in the area of surface water management, the agency: (1) grants permits for special fishing, for commercial fishing in inland waters, except in private water bodies, coordinates draft decisions granting a right to use a fishing area and plans for measures aimed at restoring and protecting fish stocks in water bodies, grants permits for using a fishing area, permits for trade in fishing equipment other than that used for recreation, permits for purchasing fishing equipment other than that used for recreation, makes decisions concerning such matters, changes, suspension and rescinding licences for such activities and approves fishing logs concerning fishing equipment records; (2) approves rules for use and maintenance of ponds in cases provided for in the law; (3) grants permits for lowering a water table level in ponds and flooded lakes; (4) approves, reviews conditions for use of surface water bodies for obtaining water with stationary and mobile water intake equipment and makes decisions to renew or rescind such conditions; (5) coordinates projects, notifications and/or modifications of projects concerning water body management or rehabilitation work; (6) grants and renews permits for laboratory tests or measurements of pollution from pollution sources and/or that released to the environment (including water) and/or for laboratory tests (other than of groundwaters), conducts – within its competence – inspections of laboratories to which these permits were granted, warns about intended permits suspension, makes decisions to suspend permits, to rescind a suspension and to rescind a permit; (7) when making a strategic assessment of an environmental impact of a plan or a programme at the communal or local level, it examines, in accordance with its competence, the choice of a strategic assessment of an environmental impact of plans or programmes, documents used to determine the scope of assessment and reports of a strategic assessment of environmental impact, draws conclusions concerning those documents, and submits proposals concerning the scope of assessment; (8) performs the functions of an institution authorised by the Ministry of Environment of the Republic of Lithuania, which is responsible for environmental protection, established by the local planning act; in accordance with the procedure set out in the law, it organises and coordinates monitoring of the status of surface water bodies; (9) uses environmental monitoring data to make assessments of the surface water status; (10) receives, collects or otherwise manages data on water use, wastewater management, water bodies, and managing flood risk based on river basins in accordance with the procedure described in the law; (11) wherever provided for in the law, performs tests of water, bottom deposits, water bodies, fauna and flora and wastewater released from sources of pollution; (12) in accordance with the procedure set out in the law, it provides data and information on the status of water bodies, changes of the coasts of the Baltic Sea and the Curonian Lagoon, pollutants introduced to water bodies to domestic and international institutions and organisations, competent organs of foreign states and to other users of data and information about the environment; and (13) in accordance with the procedure set out in the law, it collects, processes, summarises and analyses statistical data concerning the use of water resources, wastewater management, prepares and distributes statistical information based on these data.

When keeping the registers, the agency performs the functions of (1) manager of state information systems, registers, and cadastres, including the cadastral manager of rivers, lakes, and ponds in the Republic of Lithuania; (2) functions of the State Register of Waste Manager; (3) functions of the manager of the information system's integrated computer system of information on the environment, and (4) functions of the manager of the system of information on environmental permits.

Communal executive organs are obliged to monitor the environmental quality, including the preparation of six-year plans. The executive organ of a commune or its authorised body submits the programme to the EPA and the Lithuanian Ministry of Environment's Geological Service – when the programme includes monitoring of groundwater and soil. This programme is approved by the Council of the Commune. Importantly, monitoring data are to be used as a source of information on the water protection issues identified in the territory of the commune when such information is provided to the authorities responsible for preparing plans and programmes of action for river basin management. The EPA, under authorisation from the Ministry of the Environment, conducts the monitoring at the national level. Plans of river basin management are drawn up for individual river basins, which contain a valid description of the status of the river basin and the factors that affect it; results of the analysis of the human impact; information on the protection of specific waters, isolated water bodies of risk, whose good status will not be achieved until 2021; and information on the measures which have to be implemented to protect waters. The competence of local authorities which arrange implementation of the environmental protection acts and regulations adopted by the government and by the Ministry of Environment include (1) management, use, and protection of allotted natural resources and environmental objects; (2) separation of the state natural resources within the limits established for communes; and (3) preparing, approving and implementing communal programmes and other environmental protection measures in order to protect and use natural resources.

## Summary

The scope of surface water protection in the Republic of Lithuania from implementation of the provisions of the WFD. The specific objectives include (1) to define criteria for effective and correct legal implementation, (2) the method for meeting the criteria of the directive in Lithuania, and (3) to assess the practical consequences of implementation in local planning. Lithuania decided that the goal of initiating the implementation of all required supplementary measures and achieving a good status in 82% of surface water bodies by 2021 (at the beginning of the second cycle of management) was 52% of water bodies with good status. A deadline extension for technical reasons was considered only for lakes and transitional and coastal waters, where more time is needed to achieve a good status or where such a good status cannot be achieved because of the natural conditions.

According to data provided in documents (2022, No. 1292) and an assessment made by scientists and experts on the status of surface water bodies (based on data collected during state monitoring in 2014–2019), it was shown that the status of surface waters is not satisfactory, because during the third period of UBR management, 64% out of the 1,193 water bodies (63% out of the 826 rivers, 64% of the 361 lakes, all of the four transitional and two coastal categories of water bodies – 100%) were classified as endangered water bodies and they failed to meet the criteria for good ecological status. The following water bodies did not meet the criteria for good ecological status: the Neman UBR (57% of the rivers and 63% of lakes), the Lielupe UBR (94% and 95%), the Venta UBR (66% and 90%), and the Daugava UBR (22% and 44% of the rivers and lakes, respectively). During the second period of UBR management, among 21 homogeneous parts of water bodies, 185% were classified as endangered parts of water bodies and failed to meet the criteria for good ecological status (48 homogeneous parts of rivers (822%), 51 homogeneous parts of lakes (357%), all 40 transitional, and 4 coastal water bodies; see Česonienė et al., 2020).

It is pointed out in the Lithuanian literature that a division of competence between individual organs that deal with surface water protection in its broad sense is regarded as imprecise. The distribution of tasks among various bodies, especially the potential role of the district board of the river basin, is not precise, which can lead to an incoherent decision-making process.

## References

Česonienė, L., D. Šileikienė, and M. Dapkienė (2020). Relationship between the Water Quality Elements of Water Bodies and the Hydrometric Parameters: Case Study in Lithuania. *Water* 12 (2):500. https://doi.org/10.3390/w12020500

Dvarioniene, J., and Z. Stasiskiene (2007). Integrated water resource management model for process industry in Lithuania. *Journal of Cleaner Production* 15 (10):950–957. https://doi.org/10.1016/j.jclepro.2006.01.009

Environmental Protection Agency's website (2024, October 18). https://aaa.lrv.lt/lt/veiklos-sritys/vanduo/upes-ezerai-ir-tvenkiniai/valstybinis-upiu-ezeru-ir-tvenkiniu-monitoringas/upiu-monitoringo-rezultatai; Lakes – https://aaa.lrv.lt/lt/veiklos-sritys/vanduo/upes-ezerai-ir-tvenkiniai/valstybinis-upiu-ezeru-ir-tvenkiniu-monitoringas

Klimaszewska, A., M. Mariański, K. Warylewska, and J. Zięty (2017). *Spółka z ograniczoną odpowiedzialnością (société à responsabilité limitée) we francuskim kodeksie handlowym*. UWM Press.

Pietkiewicz, M., A. Klimach, and M. Ogryzek (2020). Legal status of surface waters — comparative study on the example of lakes. *Water* 12:326. https://doi.org/10.3390/w12020326

Plunge, S., M. Gudas, and A. Povilaitis (2022). Expected climate change impacts on surface water bodies in Lithuania. *Ecohydrology & Hydrobiology* 22 (2):246–268. https://doi.org/10.1016/j.ecohyd.2021.11.004

Plunge, S., M. Gudas, A. Povilaitis, and M. Piniewski (2023). Evaluation of the costs of agricultural diffuse water pollution abatement in the context of Lithuania's water protection goals and climate change. *Environmental Management* 71 (4):755–772. https://doi.org/10.1007/s00267-022-01745-1

Plunge, S., and M. Piniewski (2024). From a test case to a trusted tool: Lithuania's evolving SWATsystem for water and agricultural management. *Scientific Review Engineering and Environmental Sciences* 33 (2):115–130. https://doi.org/10.22630/srees.9790

Zębek, E. (2018). Legal protection of waters in the context of human rights. *Themis Polska Nova* 1 (13):14–24. https://doi.org/10.15804/tpn2018.1.02

# 6 Legal protection of surface waters in Germany

*Claudia Conradi and Janusz Orłowski*

## Introduction

Provisions of the Water Framework Directive (WFD) were implemented into the German law in Gesetz zur Ordnung des Wasserhaushalts (*Wasserhaushaltsgesetz* – WHG) of 31 July 2009, lately amended by Article 7 of the Act of 22 December 2023. The requirement of sustainable water use from Article 1b WFD is reflected in § 1 WHG, according to which the law pursues the purpose of 'sustainable water management' (cf. Drost, § 1 E.n. 9 (§ 1, 4)).

The Constitution of the Federal Republic of Germany, called Grundgesetz (GG) of 23 May 1949, lately amended by Article 1 des Gesetzes zur Änderung des Grundgesetzes of 19 December 2022 states in Article 30 GG that: 'Exercising the rights of the state and performance of the state tasks is the responsibility of the individual federal states unless the Constitution states otherwise and allows for different legal regulations'. This basic rule is a positive expression of the state quality of the individual federal states (cf. Degenhart, 171). According to Article 31, federal law takes precedence over individual state law. Consequently, Article 31 GG regulates the resolution of conflicts between federal law and individual state law (cf. Kment & Jarass at Jarass/Pieroth/Kment, Article 31 E. n. 1, 691). According to Article 70 (1) and (2), the individual federal states have legislative powers unless the Constitution imparts such authority to the federal state. Article 70 GG is the concretisation of Article 30 GG for the area of legislative powers (cf. Degenhart, 171). The federal state only has the legislative powers assigned to it (cf. Kment at Jarass/Pieroth/Kment, Article 70 E. n. 1, 854). Following up on this, Articles 71, 73 GG state the exclusive legislation and Articles 72, 74 GG the competitive legislation of the federal state (cf. Degenhart, 56). The competitive legislation of the federal state also covers water management in Articles 72 (1), 74 (1) n. 32 GG. Competitive legislation means the individual federal states may only enact a law as long as, and to the extent that, the federal state has not made use of its legislative priority and when they are permitted to enact laws 'at variance' with federal law under Article 72 (3) GG (cf. Schnapauff/von Knobloch at Hömig/Wolff, Article 72 E. n. 2, 559).

DOI: 10.4324/9781003586913-7

In the legislative area covered by Article 72 (3) GG the individual federal states are permitted to pass state legislation at variance with federal law. Article 72 (3) GG supersedes the more general rule in Article 31 GG (cf. Schnapauff/von Knobloch at Hömig/Wolff, Article 72 E. n. 4, 561, 562). In the event of a conflict between federal and state law, Article 72 (3), 3 GG stipulates that the later law shall take precedence (cf. Kment at Jarass/Pieroth/Kment, Article 72 E. n. 32.). According to Article 72 (3) n. 5 GG states may enact law concerning water balance, except substance- or system-related. The 'variance competence' is, when it comes to water, the reason for valid acts in each of the 16 individual federal states, and at the same time, federal laws, taking into account the WFD, are in force (cf. Czychowski & Reinhardt, Intro. E. n. 38, 17, 18). Due to the exceptions regulated in Article 72 (3) n. 5 GG in favour of the federal state, substance, or system-related regulations of the water balance are core competences of the federal government (cf. Kment at Jarass/Pieroth/Kment, Article 72 E. n. 4, 867). This is the reason why issues related to wastewater management, and, in particular, the chemical composition and the basic technical rules of wastewater discharge to waters, are basically dealt with in the federal act and the executive regulation for it (cf. Czychowski & Reinhardt, Intro. E. n. 39, 18, 19).

Cooperation between the federal state and the individual state governments is institutionalised in the Federal/States Working Group on Water (*Bund/Länder-Arbeitsgemeinschaft Wasser* – LAWA), which also develops technical working materials for the implementation of the WFD (cf. (28 February 2024) https://www.lawa.de/Publikationen-363-Wasserrahmenrichtlinie.html). Despite the joint national responsibility of the individual federal state and the federal state, the federal government remains obliged to realise the objectives of the WFD in relation to the EU (cf. Czychowski & Reinhardt, Intro. E. n. 76, 38, 39).

According to Article 83 GG, implementing the federal acts is the responsibility of the individual federal states unless the Constitution states otherwise and allows for different legal regulations. Consequently, Article 83 GG is the continuation of the basic rule of Article 30 GG for the area of administration (cf. Degenhart, 171). The individual federal states implement the WHG as a separate matter in accordance with Articles 83, 84 GG; they are the addressees of the management obligation under water law (cf. Czychowski & Reinhardt, Intro. E. n. 40, 19, 20). Based on Article 83 GG, the Constitution makes a different provision for the administration of federal water routes. In this respect, the federal government is responsible for administration in accordance with Article 89 GG.

The structure of the WHG proceeds from the general to the specific, and the standards are arranged one after the other.

Chapter 1 contains general provisions, defines the purpose of the Act (§ 1 WHG) and its scope of application (§ 2 WHG). Section 3 defines terms that are used repeatedly in the act in order to ensure a uniform understanding. In addition, § 4 makes fundamental statements on restrictions on property through water law regulations, and § 5 regulates general duties of care towards everyone in the event of a possible impact on water bodies.

Chapter 2 regulates the management of water bodies and lays down common provisions for this in Section 1 (§§ 6-24), before specific regulations on surface waters in Section 2 (§§ 25–42), coastal waters in Section 3 and 3a (§§ 43–45l) and groundwater in Section 4 (§§ 46–49) follow in the subsequent sections.

Chapter 3 contains special water management provisions on wastewater disposal in Section 2 (§§ 54-61), the handling of substances hazardous to water in Section 3 (§§ 62-63), water protection officers in Section 4 (§§ 64-66), and liability for changes to water bodies in Section 8 (§§ 89, 90).

Chapter 4 regulates compensation and pre-emption rights, Chapter 5 regulates water supervision, and Chapter 6 regulates fines and transfer provisions.

In the following sections, some important regulations for the management of surface waters, wastewater disposal, and the handling of substances hazardous to water will be examined in more detail.

### Federal legal basis of surface water management and protection

The provisions of WHG established the main rules of long-term management of waters, which are 'part of the natural environment as the foundation of human life and the living space for animals and plants, as well as protection of useful good' (§ 1 WHG). This act is applicable to surface, coastal, and groundwaters (§ 2 WHG). Surface waters (*oberirdische Gewässer*) are defined as waters that are permanent or temporarily flowing in beds or standing or waters flowing out of springs (§ 3 n. 1 WHG).

Firstly, the general principles of water management in accordance with § 6 WHG, which apply to all forms of water management, should be emphasised. In simplified terms, this is as follows. Water bodies must be managed sustainably, in particular with the aim to achieve the following:

1 Maintain and improve their functionality and performance as a component of the ecosystem and as a habitat for animals and plants, in particular by protecting them from detrimental changes in watercourse characteristics; adverse effects on the water balance must be avoided for the common good and in harmony with the interests of individuals, and to maintain and create utilisation options for the public water supply.
2 Prevent possible consequences of climate change to ensure natural and harmless runoff conditions in surface waters as far as possible.

In addition, water bodies that are in a natural or near-natural state should remain in this state. Waters that have not been restored to a near-natural state should be returned to a near-natural state.

In accordance with § 7 (1), WHG water bodies are to be managed on a river basin basis, i.e., from the source to the estuary with all tributaries.

The decisive factor is therefore no longer national and state borders, but the borders of the hydrological catchment areas. This fulfils the mandate of Article 3 (1) 1 WFD (cf. Drost, § 7 E. n. 6 (§ 7, 5). The term 'river basin district' is legally defined in § 3 n. 15 WHG as the main unit for the management of catchment areas, which consists of one or more neighbouring catchment areas, the groundwater attributable to them, and the coastal waters belonging to them. Section 7 (1) WHG lists a total of ten river basin districts by name within the scope of the WHG (Danube, Rhine, Meuse, Ems, Weser, Elbe, Eider, Oder, Schlei/Trave, and Warnow/Peene). The river basin districts are shown on maps in Annex 2 to the WHG.

Section 7 (2) WHG obliges the federal states that are part of a river basin district to coordinate their water management planning and measures (cf. Czychowski & Reinhardt, § 7 E. n. 8, 153). Across borders, § 7 (3) WHG aims to harmonise administrative activities with the authorities of other member states and to coordinate with the administrations of affected foreign states that are not members of the EU. For the river Oder, for example, the three countries involved in the river catchment area – Germany, Czech Republic, and Poland – have concluded an international treaty on the International Commission for the Protection of the Oder against Pollution (shortened ICPO) for the purpose of coordination at international level (cf. Foundation contract BGBl. 1997 II S. 1708; Amendment contract BGBl. 2010 Teil II No. 25, 1054). The aims of the ICPO are

to prevent and reduce the pollution of the Oder and the Baltic Sea with pollutants;

to achieve near-natural aquatic and associated terrestrial ecosystems with a corresponding diversity of species;

to enable the utilisation of the Oder, especially for the production of drinking water from bank filtrate and the agricultural use of the water and sediments;

to prevent or sustainably reduce the risk of flood damage; and

to coordinate the implementation of the WFD in the Oder catchment area (cf. International Commission for the Protection of the Odra River against Pollution (25 March 2024), https://www.mkoo.pl/index.php?mid=1&lang=EN).

Comparable agreements were concluded for all other river basin districts in accordance with § 7 (1) WHG. The core task of the international commissions is to identify the regulatory problems existing between the countries involved in the water sector and to develop solutions in order to realise the common interests of the riparian states in the use and protection of rivers (cf. International Commission for the Protection of the Odra River against Pollution (25 March 2024), https://www.mkoo.pl/index.php?mid=1&lang=EN).

Section 7 (4) WHG requires coordination with the competent federal authorities if federal waterways or transboundary river basin districts are the subject of planning. Section 7 (5) WHG assigns the federal states the task of allocating the respective catchment areas as well as coastal waters and groundwater within their state borders to a river basin district under § 7 (1) WHG.

Section 7 WHG primarily serves to implement the WFD. This means that the federal states cannot deviate from it. As part of the implementation of the WFD, the federal states have issued all supplementary regulations to § 7 WHG (cf. Drost, § 7 E. n. 12 (§ 7, 9).

According to § 8 WHG, the use of a body of water requires a permit or authorisation, unless otherwise stipulated by the Water Resources Act. Under certain conditions, the use of waters for emergency response and defence purposes is exempt from the permit and authorisation requirement.

Water use includes, according to § 9 WHG,

1 intake and channelling away water from surface waters;
2 impounding and draining surface waters;
3 collecting solids from surface waters, if this results from the water properties;
4 introducing substances to waters; and
5 intake and channelling away groundwaters.

Moreover, water use includes diverting groundwaters to installations intended for this purpose and opening devices with the use of hydraulic pressure to search for natural gas, oil or geothermal energy.

If the use of waters requires a permit or authorisation, this must be refused under the conditions of § 12 WHG if

1 harmful changes to water bodies are to be expected that cannot be avoided or compensated for by ancillary provisions, or
2 other requirements under public law are not met.

Otherwise, according to § 12 (2) WHG, the granting of the permit and authorisation is at the discretion of the authority (management discretion).

Section 12 WHG conclusively regulates the requirements for the granting of a permit and authorisation. There is no recognisable scope for supplementary regulations by the federal states within the framework of Article 72 (1) GG (cf. Berendes, § 12 E. n. 8, 100).

The federal law does not establish the competences of the water authorities in these issues because, as already mentioned, the implementation of the WHG is the responsibility of the individual federal states in accordance with Articles 83, 84 of the Constitution.

Based on these exemplary common provisions for water management, the following section will provide insight into the basic provisions of the WHG on the management of surface waters. The management of surface waters is regulated in §§ 25–42 WHG.

First of all, §§ 25 and 26 WHG regulate exceptions to the principle of authorisation and approval requirements for surface waters in deviation from § 8 WHG.

Section 25 WHG makes an exception '[...] for almost exclusively traditional, (water management) less important types of use' (cf. Czychowski & Reinhardt, § 25 E. n. 2, 461), in which public use is permitted. Public use is the authorisation granted to any person under public law to use a public object without special permission within the scope of its sovereign purpose (cf. Drost, § 25 E. n. 11 (§ 25, 6). According to § 25 WHG, the federal states are responsible for authorising public use. Authorisation is permitted provided that it does not conflict with the rights of others and does not impair other rights or the use of other owners or residents.

Section 26 WHG allows owners and residents to use surface waters for their own needs without a permit or authorisation under certain conditions if, among other things, no adverse changes to the water quality are to be expected.

With regard to water management in §§ 27–31 WHG, the basic management concept of the WFD for surface waters can be found. Section 27 WHG is particularly important, as it defines the management objectives for surface waters.

According to § 27 (1) WHG, surface waters, unless they are classified as artificial or heavily modified in accordance with § 28 WHG, must be managed in such a way that

1 a deterioration of their ecological and chemical status is avoided and
2 good ecological and good chemical status can be maintained or achieved.

According to § 27 (2) WHG, water bodies within the scope of § 28 WHG must be managed in such a way that

1 a deterioration of their ecological potential and chemical status is avoided and
2 good ecological potential and good chemical status are maintained or achieved.

Section 27 WHG formulates the general principle for the management of surface waters; from an ecological and chemical perspective, it contains both the prohibition of deterioration and the requirement to maintain or achieve good status or good potential (cf. Czychowski & Reinhardt, § 27 E. n. 5, 486).

Consequently, the provision in § 27 WHG, as a 'guiding provision', opens up the series of water management regulations for implementing the environmental objectives under Article 4 WFD in federal law (cf. Czychowski & Reinhardt, § 27 E. n. 2, 484). The requirements of § 27 WHG are directly applicable law that must be observed by the authority for every project authorisation under water law (cf. Czychowski & Reinhardt, § 27 E. n. 5, 486).

In practice, the prohibition of deterioration in § 27 (1) WHG is particularly important.

In the absence of a legal definition of the term 'deterioration', it was initially disputed in case law and doctrine as to when a 'deterioration' within the meaning of § 27 WHG exists.

In its decision on the deepening of the River Weser, the European Court of Justice (shortened ECJ) clarified that, in direct reference to the wording of Article 4 (1) WFD, deterioration requires an actual, demonstrable risk of deterioration (cf. ECJ, judgement of 1 June 2025 – C-461/13, NVwZ 2015, 1042–1047). According to the ECJ, this is to be assumed as soon as the status of at least one quality element within the meaning of Annex V WFD deteriorates by one class, even if this does not lead to a deterioration in the classification of the water body as a whole.

In view of the extensive requirements of EU law, the WHG has shifted the regulation of important detailed issues relating to the management of surface waters to the ordinance level. According to § 23 (1) n. 1 to 3 and 8 to 12, as well as (2) WHG, special requirements for the quality of water bodies, the utilisation of water bodies, and the determination, description, delimitation, classification, and presentation of the status of water bodies are to be regulated by ordinance. The same applies to the monitoring of water characteristics, the requirements for measurement methods and procedures, and economic analyses.

Based on this authorisation, the Ordinance on the Protection of Surface Waters (Verordnung zum Schutz von Oberflächengewässern – Oberflächengewässerverordnung – OGewV vom 20 June 2016, lately amended by Article 2 (4) des Gesetzes vom 9 December 2020) regulates the details of surface water protection nationwide and contains regulations on the classification, typification, and delineation of surface water bodies in accordance with the requirements of the WFD.

The provisions on the programme of measures (§ 82 WHG) and the management plan (§ 83 WHG) are to be seen in connection with the environmental objectives set out in § 27 WHG.

Section 82 WHG transforms Article 11 WFD into federal law, and in § 82 (1), WHG obliges the authorities responsible under federal state law to draw up programmes of measures for each river basin district to achieve the management objectives of §§ 27–31, 44, 47 WHG (cf. Berendes, § 82 E. n. 1, 464).

In implementing this regulation, the federal states are responsible for drawing up and updating the plans (cf. Czychowski & Reinhardt, § 82 E. n. 4, 1245). In addition, the federal states remain free to add normative amendments (in particular procedural details), including the option of declaring the programmes binding (cf. e.g., § 87 SächsWG, § 51 BayWG, § 24 BdgWG, § 130a MecklVorpWG; Czychowski & Reinhardt, § 82 E. n. 4 wfe., 1245).

The programme of measures is thus the link between the abstract management objectives in Section 27 (1) WHG (for natural water bodies) and individual decisions by the water authority (Queitsch at Wellmann, Queitsch & Fröhlich, § 82 E. n. 2, 657). Section 82 WHG contains binding specifications on the content of the programme of measures. In addition to the WFD itself, documents from the EU Commission, the cross-river basin work of the LAWA, and specifications within the framework of national and international coordination within the river basin district form the main basis for the development of the programme of measures.

According to § 83 WHG, a management plan must be drawn up for each river basin district.

§ 83 (2) WHG concretises this obligation by stipulating that the contents of Article 13 (IV) in conjunction with Annex VII of the WFD must be included in the plans (cf. Czychowski & Reinhardt, § 83 E. n. 12, 1267). The management plan is an important instrument for implementing the WFD, as it integrates all the information required by the WFD for river basin-related water management with the aim of achieving good ecological status or good ecological potential and good chemical status for surface waters and good quantitative and chemical status for groundwater.

The highest water authorities of the individual federal states are generally responsible for implementing the programme of measures in accordance with § 82 WHG and for drawing up and updating the management plan in accordance with § 83 WHG, which is regulated accordingly in the state water laws (cf. § 87 (2) SächsWG, Article 51 (2) BayWG, § 13 (2) WG Baden-Würtemberg).

They coordinate and monitor the implementation of the measures by private and/or public organisations in their local area of responsibility.

The programme of measures (§ 82 WHG) and the management plan (§ 83 WHG) adopted by the federal states are published and are therefore binding for the authorities. Although the action plan and management plan are only internal administrative plans (cf. to the strike stand: Czychowski & Reinhardt, § 82 E. n. 10-12aff., 1247–1250 ff. and § 83 E. n. 8, 8a, 1265–1266), they are the central instruments for controlling and coordinating integrated water management in the EU member states (cf. Berendes, § 82 E. n. 5, 466 and § 83 E. n. 6, 475).

In addition to the provisions of the WHG and the OGewV, the following federal ordinances are important for the protection of surface waters:

- Verordnung über Anforderungen an das Einleiten von Abwasser in Gewässer (Abwasserverordnung – AbwV) in der Fassung der Bekanntmachung vom 17 June 2004, lately amended by Article 1 der Verordnung vom 20 January 2022 and
- Verordnung über Anlagen zum Umgang mit wassergefährdenden Stoffen (AwSV) of 18 April 2017, lately amended by Verordnung of 19 June 2020

The latter regulates the specific technical requirements for the operation of facilities for handling substances hazardous to water, including notification and suitability determination procedures. In this respect, the AwSV substantiates the general provisions of §§ 62 and 63 WHG. Facilities for handling substances hazardous to water must be designed, constructed, maintained, operated and decommissioned in such a way that there is no reason to fear any adverse change in the properties of watercourses.

### Legal provisions of surface water management and protection in the individual federal states

The Federal Water Act (WHG) regulates the fundamental, nationwide issues relating to the utilisation of water resources and the management of these matters. Due to the concurrent legislation of the federal government and the federal states, three case constellations must be distinguished with regard to the legislation of the federal states:

Case 1: The federal states can supplement federal water resources legislation in areas that are not conclusively regulated.
Case 2: The WHG or federal law can authorise the federal states to 'determine otherwise' or make 'deviating regulations'.
Case 3: The federal states can deviate from federal regulation and the WHG within the limits of Article 72 (3) GG.

However, EU law (supranational law) and intergovernmental agreements (international law) are supreme.

The following state laws and the implementing ordinances based on them can be found on the websites of the state governments.

**Baden-Württemberg:** Wassergesetz für Baden-Württemberg (WG) of 3 December 2013, lately amended by Article 9 der Gesetzes of 7 February 2023. For further legal regulations up to administrative regulations see also (25 March 2024) https://um.baden-wuerttemberg.de/de/umwelt-natur/wasser/rechtsvorschriften.

**Bavaria:** Bayerisches Wassergesetz (BayWG) of 25 February 2010, BayRS 753-1-U), lately amended by § 1 des Gesetzes of 9 November 2021. For additional legal bases, see also (25 March 2024) https://www.stmuv.bayern.de/themen/wasserwirtschaft/recht/index.htm.

**Berlin:** Wassergesetz (BWG) as amended by the announcement of 17 June 2005, lately amended by Article 2 des Gesetzes of 25 September 2019. For additional legal bases, see also (25 March 2024) https://www.berlin.de/sen/uvk/service/rechtsvorschriften/umwelt/wasser-und-geologie/.

**Brandenburg:** Brandenburgisches Wassergesetz (BbgWG) as amended by the announcement of 2 March 2012, lately amended by Article 1 des Gesetzes of 4 December 2017. For further legal regulations up to administrative regulations, see also (25 March 2024) https://mluk.brandenburg.de/mluk/de/service/rechtsvorschriften/wasser/.

**Bremen:** Bremisches Wassergesetz (BremWG) of 12 April 2011, lately amended by Article 6 n. 5 des Gesetzes of 24 November 2020.

**Hamburg:** Hamburgisches Wassergesetz (HWaG) as amended by the announcement of 29 March 2005, lately amended by Article 12 des Gesetzes of 4 December 2012.

**Hesse:** Hessisches Wassergesetz (HWG) vom 14 December 2010, lately amended by Article 39 des Gesetzes of 28 June 2023 (GVBl. S. 473, 475).

**Mecklenburg-Western Pomerania:** Wassergesetz des Landes Mecklenburg-Vorpommern (LWaG) of 30 November 1992, lately amended by Gesetz of 8 July 2021.

**Lower Saxony:** Niedersächsisches Wassergesetz (NWG) of 19 February 2010, lately amended by § 2 WasserG-Bagatellgrenze-ÄndVO vom 6 December 2023.

**Rhineland-Palatinate:** Landeswassergesetz (LWG) of 14 July 2015, lately amended by Article 2 des Gesetzes of 8 April 2022.

**North Rhine-Westphalia:** Wassergesetz für das Land Nordrhein-Westfalen (Landeswassergesetz – LWG) as amended by the announcement of 8 July 2016 lately amended by Article 3 des Gesetzes of 17 December 2021 (GV. NRW. S. 1470).

**Saarland:** Saarländisches Wassergesetz (SWG) as amended by the announcement of 30 July 2004, lately amended by Article 173 des Gesetzes of 8 December 2021. For further regulations, see also (26 March 2024) https://www.lanuv.nrw.de/umwelt/wasser/service/gesetze-regelwerke-und-mehr.

**Saxony:** Sächsisches Wassergesetz (SächsWG) of 12 July 2013, lately amended by Article 12 des Gesetzes of 20 December 2022.

**Saxony-Anhalt:** Wassergesetz für das Land Sachsen-Anhalt (WG LSA) of 16 March 2011, lately amended by Article 21 des Gesetzes of 7 July 2020.

**Schleswig-Holstein:** Landeswassergesetz (LWG) of 13 November 2019, § 101 amended by Article 3 n. 3 Gesetz. of 6 December 2022. For further regulations, see also (15 October 2024)

https://www.schleswig-holstein.de/DE/fachinhalte/W/wasserwirtschaft/rechtlicheGrundlagen?nn=3cf9b76a-f91e-4d3f-ba72-0142e3ad0cf9

**Thuringia:** Thüringer Wassergesetz (ThürWG) of 28 May 2019; lately amended by Article 17 des Gesetzes of 11 June 2020.

Provisions of the state laws concerning the water law (*Wassergesetze*) are complemented – as has been pointed out earlier – with numerous and practically important regulations (*Verordnungen*) and so-called internal administrative regulations (*Verwaltungsvorschriften*) (cf. Berendes, Intro. E. n. 33, 19).

German water law is subordinate to the laws of the EU (supranational law) and complements its norms. International agreements play a similar role, especially when issues related to environmental protection reach beyond state borders (cf. Berendes, Intro. E. n. 34–36, 19, 20).

**Responsible water authority in individual federal states**

The question of which authority in each federal state is responsible for implementing the water protection regulations cannot be answered uniformly but depends on how the administrative structure is organised.

The administrative structure of the individual federal states is organised by themselves, so there are certainly differences. Nevertheless, general principles in the administrative structure of the federal states can be described. The following four different levels can be identified:

Supreme state authorities /highest-level authority *(Oberste Landesbehörden)*:
State government, usually consisting of the State Chancellery and 8 to 12 ministries.
Higher state authority *(Obere Landesbehörden)*:
Higher state authorities, state enterprises
Middle state authorities *(Mittlere Landesbehörden)*:
Specialist resource instances
Lower state authorities *(Untere Landesbehörden)*:
Subordinate authorities and subordinate organisations (districts, independent cities, communities, towns, and communities belonging to the district)

There is no middle state authority in the states of Brandenburg, Mecklenburg-Western Pomerania, Lower Saxony, Saarland, and Schleswig-Holstein (cf. Czychowski, 113–116).

The following are public administration bodies competent in matters related to water management:

**Baden-Württemberg:** Ministry of Environment, Climate Protection and the Energy Sector as the supreme administrative authority, Government Praesidia as organs of a higher level, organs of a lower level are district offices *(Landratsämter)* and communities *(Gemeinden)* (§ 15 Landesverwaltungsgesetz (VwG BW) vom 14 October 2008, lately amended by Article 10 des Gesetzes of 21 May 2019. Following on from that, § 82 WG BW regulates the substantive jurisdiction. Accordingly, the lower water authority is responsible unless the law stipulates otherwise, § 82 (1) WG BW.

**Bavaria:** The Bavarian State Ministry for the Environment and Consumer Protection is the supreme administrative authority. Part 5 of the BayWG regulates water supervision. Article 58 BayWG stipulates that the district administrative authorities are responsible for water supervision in accordance with § 100 WHG. They order the measures that are necessary in individual cases at their own discretion. Technical water supervision is the responsibility of the specialised authorities subordinate to the State Ministry. Article 63 BayWG regulates the responsibility of the authorities, both for the enforcement of the WHG in general and for the implementation of the BayWG and ordinances issued on its basis. Accordingly, the district administrative authorities are also responsible for enforcement, unless otherwise specified For further information on the organisation of the authorities (23 February 2024), see https://www.stmuv.bayern.de/themen/wasserwirtschaft/wasserwirtschaft_in_bayern/organisation.htm

**Berlin:** According to § 85 BWG supervision in water management is the responsibility of the Senate administration (*Senatsverwaltung*) and regional water offices (*Bezirksämter*). A district office is competent in matters related to damage in water management and granting permits and supervision over wastewater installations.

**Brandenburg:** the Ministry of Agriculture, Environment and Climate Protection is competent in water management, the highest-level authority and the state office for the natural environment (*Wasserwirtschaftsamt des Landes Brandenburg*) is a higher-level authority. Country districts and independent cities are lower-level bodies. In accordance with § 126 BbgWG, supervision in water management is – in principle – the responsibility of lower-level-bodies. Special supervision over country districts is exercised by the highest-level body.

**Bremen:** According to § 92 BremWG, the following water administration bodies in Bremen are Senator for Climate Protection, Environment, Mobility, Urban Development and Construction of Bremen and the coastal area of Bremerhaven (as the higher-level body) and the magistrate of the city of Bremerhaven for the urban commune of Bremerhaven (corresponding to the lower water authorities in other federal states) This Senator can issue regulations to set the competence of other administrative bodies for matters related to water.

**Hamburg:** The Arrangement of responsibilities in the field of water law and water management (*Anordnung der Zuständigkeiten auf dem Gebiet des Wasserrechts und der Wasserwirtschaft* – WasRZustAnO HA vom 7 April 1987, lately amended by Article 143 der Anordnung vom 6 October 2020) assigns responsibility for water law and water management in Section I (1) to the Ministry of the Environment, Climate, Energy and Agriculture. It is responsible for the implementation of WHG and the HWaG. In particular, it is responsible for the fulfilment of ministerial tasks and is therefore the highest administrative authority. It has a subordinate Office for Water, Wastewater, and Geology, divided into departments (Water, Wastewater, and Geology, Conservation of Nature, Planning of Green Areas and Soil Protection, Protection against Emissions and Waste Management, Energy and Climate, and Agriculture).

The tasks of the lower water authority are carried out by the district offices. In accordance with Section II WasRZustAnO HA, they are responsible, for example, for tasks relating to surface waters, including water supervision. Section III WasRZustAnO HA regulates special responsibilities, for example, the Hamburg Port Authority for the harbour area and storm surge protection. Tasks and power of Hamburg´s water administration body are regulated in §§ 64 ff. HWaG.

**Hesse:** According to § 64 HWG, the Ministry for Water Management is the highest-level office in matters related to water in Hesse. The government praesidium is a higher-level body. The tasks assigned to a lower-level body are performed by a district commission and the magistrate of an independent city. Moreover, specific competences in water-related matters are also given to the Hesse Office for Conservation of Nature, Environment and Geology.

**Lower Saxony:** According to § 127 NWG, the ministry with subject-matter competence (*Fachministerium*) is the highest-level office in Lower Saxony. District boards are higher-level offices. Districts, large and independent cities, are the lowest-level bodies. They can conclude agreements for task performance upon the government´s consent. The responsibilities of the individual authorities are regulated at Verordnung über Zuständigkeiten auf dem Gebiet des Wasserrechts (ZustVO-Wasser) of 10 March 2022, lately amended by Article 6 des Gesetzes of 19 December 2010. Available at (23 February 2024) https://voris.wolterskluwer-online.de/browse/document/dd3825fb-ee8c-3b00-a9c5-9d35ce7b9025.

**Mecklenburg—Western Pomerania:** According to § 106 LWaG, the Ministry of Agriculture, Environment, and Consumer Protection is the highest-level body for water-related matters in Mecklenburg-Western Pomerania. The State Office for Environment, Nature Conservation, and Geology is the higher-level water administration body. Lower water authorities are state agriculture and environment offices, as well as district heads and mayors of independent cities. They are, in principle, responsible for enforcement of the law.

**Rhineland-Palatinate:** The Ministry for Climate Protection, Environment, Energy, and Mobility is the highest-level office in water-related matters. The Structure and Permit Directorate is a higher-level body, and the administration of districts, independent cities, and cities are lower-level bodies (§ 92 LWG). Section 19 LWG regulates the responsibilities of the higher and lower water authorities, among other things, for issuing the relevant licences and coordinating the use of the waters. For the discharge of wastewater from households and similar wastewater, responsibility is based on the volume of wastewater treated. Up to 8 $m^3$ per day, the lower water authority is responsible, and from 8 $m^3$ per day, the responsibility is that of the higher water authority. The North and South Structural and Licensing Directorates as the higher water authorities, and the district and independent city administrations as the lower water authorities.

**North Rhine-Westphalia:** The Ministry for the Environment, Nature, and Transport is the highest-level body. The regional board is a higher-level administration organ, and the district and independent city are lower-level ones (§ 114 LWG).

**Saarland:** According to § 102 SWG, the Ministry of Environment and Consumer Protection is the highest-level body for water-related matters.

The following are lower-level bodies: The State Office for Environment and Labour Protection and lower construction supervision bodies. Section 103 SWG regulates the substantive jurisdiction of the water authorities in detail, and § 104 SWG regulates the local jurisdiction of the lower construction supervision bodies.

**Saxony:** According to § 109 (1) SächsWG, the following are water administration bodies with general competence: (1) State Ministry (*Staatsministerium*) of Environment and Agriculture – as the highest-level water body, (2) State Directorate of Saxony as a higher-level body, (3) districts and independent cities as the lower-level water organs. According to § 109 (2) SächsWG, the following are special bodies:

The State Office for Environment, Agriculture, and Geology as a specialised technical body for professional advising and support for the supreme water authority and (2) the National Board for Water Engineering Structures – State Company (as a unit for water structures). Section 110 (1) SächsWG regulates the substantive jurisdiction of the water authorities. Accordingly, the lower water authority is responsible for enforcing the WHG and the SächsWG, as well as the legal ordinances issued on the basis thereof, unless otherwise specified. According to § 110 (2) SächsWG, the supreme water authority can regulate the responsibilities for the enforcement of water law regulations by statutory order. In accordance with § 102 (2), 2 SächsWG, the decisive criterion for the transfer of responsibility is the reliable and appropriate fulfilment of tasks, which corresponds to constitutional requirements (cf. Gläs at Dallhammer/Dammert/Faßbender, § 110 E. n. 6, 583). It has exercised this authorisation with the Saxon Water Competence OrdinanceSächsische Wasserzuständigkeitsverordnung–SächsWasserZuVO (official designation: Joint Ordinance for the Environment and Agriculture and the Saxon State Ministry for Social Affairs and Consumer Protection on responsibilities in the field of water law and water management; in German: *Gemeinsame Verordnung für Umwelt und Landwirtschaft und des Sächsischen Staatsministeriums für Soziales und Verbraucherschutz über Zuständigkeiten auf dem Gebiet des Wasserrechts und der Wasserwirtschaft*) of 12 June 2014 (SächsGVBl. S. 363), lately amended by Article 13 HaushaltsbegleitG 2023/2024 of 20 December 2022. The SächsWasserZuVO regulates the responsibilities of the supreme and higher water authorities as well as the responsibilities of the State Office for the Environment and Geology and the state-owned companies.

**Saxony-Anhalt:** According to § 10 WG LSA, the ministry with responsibility for water management, therefore the Ministry for Science, Energy, Climate Protection and the Environment, is the highest-level body. The State Administration Office is the higher-level body, and districts and independent cities are lower-level bodies. The State Office for Environmental Protection is competent in technical matters, and it provides support to higher and lower water offices in particularly complicated matters. The State Institution for Protection against Flood and for Water Management in Saxony-Anhalt performs water status analyses. Section 112 WG LSA regulates the substantive and local jurisdiction of the water authorities in detail.

**Schleswig-Holstein:** According to § 101 (1) LWG, the ministry with competence for water management is the highest-level body, the State Office of Natural Environment – the higher-level one – and district heads and mayors of independent states – the lowest-level bodies. Section 101 (2) LWG stipulates that the aforementioned ministry is authorised to rule the substantive jurisdiction of the water authorities by a separate legislative regulation. Moreover, according to § 102 (1) and (2) LWG, *there* are coastal protection organs: the Ministry for Coastal Protection is the highest-level body, and bodies designated by a separate legislative regulation are lower-level ones. The Ministry of Energy Transition, Climate Protection, Environment, and Nature is responsible for water management, as well as coastal protection, and it has, in order to implement § 101 (2) and § 102 (2) LWG, enacted the Landesverordnung über die Zuständigkeit der Wasser- und Küstenschutzbehörden (*Wasser- und Küstenschutz-Zuständigkeitsverordnung* – WaKüVO) vom 4 December 2019. This ordinance regulates in detail the areas of responsibility of the supreme state authorities, the higher state authority, the lower state authorities, as well as the authority of the coastel protection authority.

**Thuringia:** According to § 59 ThürWG, the ministry with competence for water management is the highest-level body. The Thuringian Office for Environment, Mining, and Nature Conservation is, as a technical authority (cf. § 60 ThürWG), the higher-level body. Districts and independent cities are the lowest-level bodies. Section 61 ThürWG regulates the substantive and local jurisdiction of the water authorities in detail. This list reflects the role of state administration in performing tasks related to water management and protection, which includes implementation of the federal WHG act as well as state legislation and international cooperation. Tasks at the lowest level of water authority are performed in general by commune local government, including independent cities (*Kreisfreie Städte*). It is noteworthy that specialised institutions of technical control (laboratories, water protection units) exist in the states, whose technical equipment and specialists are employed there to perform the tasks of water authorities. The state law obligates enterprises discharging wastewater into flowing waters to measure the content of substances in it.

### Federal legal provisions concerning wastewater management

According to § 54 (1) WHG, wastewater is water that has changed in its properties through use for domestic, commercial, agricultural, or special purposes and the water that runs off with it in dry weather (wastewater) as well as rainwater. Wastewater also includes liquids produced and collected from facilities for treating, depositing, and storing waste.

In accordance with 54 (2) WHG, wastewater disposal includes the collection, transport, treatment, discharge, seepage, and irrigation of wastewater, as well as the dewatering of sewage sludge in connection with wastewater disposal.

The basic rules for wastewater discharge are laid down in § 55 WHG. Wastewater is to be discharged in such a way as to 'not affect the public good' (*das Wohl der Allgemeinheit nicht beeinträchtigt wird*). The public good is the objective of the WHG and thus the decisive criterion for all measures for wastewater disposal, which is both a public service (basic supply) and, at the same time, a measure for environmental protection and epidemic defence (cf. Czychowski & Reinhardt, § 55 E. n. 7, 866). The disposal of domestic wastewater through decentralised systems can also be in the public interest, § 55 (1) WHG. However, this regulation does not give property owners a legal right to dispose of their wastewater in a decentralised wastewater system. According to the case law of the Federal Administrative Court, this option only exists until a connection to the public sewerage system is available. Once the connection has been established, the municipal obligation to connect and utilise applies (cf. BVerwG, Beschluss vom 19 December 1997 – 8 B 234/97 at NVwZ 1998, 1080–1082; DVBl. 1998, 1222–1223).

Rainwater should be infiltrated close to the site, trickled or discharged directly into a body of water via a sewerage system without mixing with wastewater, provided that this does not conflict with any public law regulations or water management concerns, § 55 (2) *WHG*. If rainwater is polluted, water protection requires that it be channelled to a sewage treatment plant via a combined sewer (cf. Queitsch at Wellmann/Queitsch/Fröhlich, § 55 E. n. 5 w.f.e, 337, 338).

Liquid substances that are not wastewater can be disposed of with wastewater if such disposal of the substances is more environmentally compatible than disposal as waste and does not conflict with water law requirements, § 55 (3) WHG. This regulation concerns a special case of wastewater disposal and is relevant, for example, for liquid, easily biodegradable production residues, which by definition do not fall under the definition of wastewater in accordance with § 54 (1) WHG but are subject to waste regulations (cf. Queitsch at Wellmann/Queitsch/Fröhlich, § 55 E. n. 38, 354). Some federal states reserve the right to make the discharge of such substances subject to authorisation in accordance with applicable state law, e.g., in North Rhine-Westphalia in accordance with § 58 (1) LWG NRW.

According to § 56 *WHG*, public law entities are obliged to wastewater removal by the provisions of the individual state law. States can set out the conditions under which other entities are obliged to discharge wastewater. These units can assign tasks in this regard to third parties. Which legal entities these are, is determined by state law (cf. Drost, § 55 E. n. 5 (§ 55, 2). Corporations under public law are primarily the municipalities and the special-purpose associations formed by them in the form of special-purpose wastewater associations, e.g., in Bavaria in accordance with Article 17 et seq.

of the Act on Municipal Cooperation (*Gesetz über die kommunale Zusammenarbeit* – KommZG vom 20 June 1994, lately amended by § 8 des Gesetzes vom 24 July 2023). Legal entities under public law also include other corporations, institutions, and foundations (cf. Queitsch at Wellmann/Queitsch/Fröhlich, § 56 E. n. 1, 356).

§ 57 WHG regulates the discharge of wastewater into bodies of water (direct [wastewater] discharge – *Direkteinleitung*) is a special regulation for the discharge of wastewater that supplements § 12 WHG (cf. Czychowski & Reinhardt, § 57 E. n. 2, 886, 887). The direct discharge of wastewater requires a permit, which may only be issued under the conditions of § 57 (1) WHG.

Accordingly, permission may only be granted if

1 the quantity and harmfulness of the wastewater is kept as low as possible in compliance with the relevant state of technical knowledge (state of the art),
2 the discharge is compatible with the requirements for water characteristics and other legal requirements, and
3 wastewater installations and other devices are constructed and used in such a way that they ensure compliance with the requirements under numbers 1 and 2.

The numbers 1 and 2 form the combined approach according to Article 10 of the WFD for the requirements for the wastewater to be discharged (cf. Drost, § 57 E. n. 8 (§ 57, 9).

According to § 57 (2) WHG, rules of wastewater removal according to technical knowledge can be set out in regulations. This regulatory power of the federal government is intended to ensure that it is legally certain which technical regulations correspond to the state of technical knowledge/state of the art. This regulation enables the adoption of supranational and international requirements into German water law (cf. Czychowski/Reinhardt, § 57 E. n. 31, p. 895).

Section 57 (2) WHG is, in particular, the legal basis for the Federal Wastewater Ordinance (*Abwasser-Verordnung des Bundes*), which specifies minimum standards within the meaning of § 57 (1) n. 1 WHG (cf. Queitsch at Wellmann/Queitsch/Fröhlich, § 57 E. n. 12, 467, 468). The Wastewater Ordinance (*Abwasser-Verordnung des Bundes*) contains 57 appendices that contain requirements for the composition of wastewater in relation to specific wastewater producers (e.g., meat industry and chemical industry).

Section 57 (1) and (2) WHG contain provisions that can be supplemented by state law within the framework of Article 72 (1) GG. Para. 1 allows for further requirements for the granting of the licence, while para. 5 can specify the necessary adaptation measures and the appropriate period. Furthermore, § 57 WHG is a substance-related regulation that does not permit any deviating state regulations in accordance with Article 72 (3) GG and is in part mandatory under European law (cf. Berendes, § 57 E. n. 14, 338).

The provisions of §§ 58 and 59 WHG regulate the rules of indirect wastewater discharge *(Indirekteinleitung)* to public and private sewage systems.

Section 58 WHG is a full regulation under federal law and defines the term 'indirect discharge' in (1) s. 1 as the discharge of wastewater into public wastewater facilities. This sets the boundary to the term 'direct discharge' in § 57 WHG (cf. Queitsch at Wellmann/Queitsch/Fröhlich, § 58 E. n. 1, 340).

According to § 58 (1), 1 WHG, the discharge of wastewater into public wastewater systems (indirect discharge) requires authorisation from the competent authority if the wastewater is subject to requirements specified in the Wastewater Ordinance (Abwasser-Verordnung des Bundes) for the place where the wastewater is produced or before it is mixed. Section 58 (1), 2 WHG authorises the federal government to replace the authorisation requirement with a notification requirement by statutory order (no. 1) and to make monitoring more flexible (no 2). These regulations serve the interests of simplifying the law and reducing bureaucracy (cf. Berendes, § 58 E. n. 3, 340, 341). According to § 58 (1), 3 and 4 WHG, further legal provisions of the federal states remain unaffected by this (cf. Queitsch at Wellmann, Queitsch & Fröhlich, § 58 E. n. 13, 485, 486).

Section 58 (2) WHG standardises the requirements to be met by a harmless indirect discharge from a federal perspective and must therefore be fulfilled as a prerequisite for the granting of a permit. According to § 58 (2), WHG authorisation for an indirect discharge may only be granted if

1 the requirements applicable to the discharge in accordance with the Wastewater Ordinance (*Abwasser-Verordnung des Bundes*) as amended, including the general requirements, are complied with;
2 the fulfilment of the requirements for the direct discharge is not jeopardised; and
3 wastewater treatment plants or other facilities are installed and operated that are necessary to ensure compliance with the requirements under numbers 1 and 2.

In this respect, § 58 WHG represents a regulation relating to both substances and facilities, which, according to Article 72 (3), 1 n. 5 GG, does not permit any deviating state law (cf. Berendes, § 58 E. n. 7, 342).

§ 59 WHG regulates the case of indirect wastewater discharge into private wastewater systems. With this provision, the federal legislator wants to take account of the increasing importance of so-called chemical and industrial parks (cf. Berendes, § 59 E. n. 1, 343). Section 59 WHG regulates the case of indirect wastewater discharge into private wastewater systems. With this provision, the federal legislator wants to take account of the increasing importance of so-called chemical and industrial parks. In principle, § 59 (1) WHG treats indirect discharges into public and private wastewater facilities equally. The same legal system therefore applies to both types of discharge (cf. Berendes, § 59 E. n. 3, 344). Due to the equal treatment of discharges in § 59 (1) WHG, reference can be made to the comments on § 58 WHG. Section 59 (2) WHG is both a substance-related and a plant-related regulation from which the federal states may not deviate in accordance with Article 72 (3), 1 n. 5 GG.

Section 60 WHG specifies the technical requirements for the construction, use, and maintenance of sewage systems, set out in accordance with the technical knowledge, the consent for their use, and provisions of the executive regulations to WHG and EU directives.

The term 'wastewater facilities' is not defined in the WHG. Wastewater facilities are all public or private facilities for the disposal of wastewater, in particular for the collection, conveyance, treatment, discharge, infiltration, sprinkling, and irrigation of wastewater and for the dewatering of sewage sludge in connection with the disposal of wastewater (cf. Czychowski & Reinhardt, § 60 E. n. 9, 923). Consequently, not only wastewater treatment plants are included but also sewers, stormwater overflow basins, drainless pits, and property drainage systems, for example.

In line with § 60 (1) – (7) WHG, sewage systems should be constructed in such a way as to meet the requirements for wastewater removal, including the widely known technical rules. If such conditions are not met, a deadline is set for a specific installation to achieve compliance with them. Installations with a specific impact on the environment require additional consent to make such alterations in their structure. If the specific requirements are not met, the relevant permit (*Genehmigung*) is denied.

The federal states can norm the rules of a device and site operation in a separate manner, i.e., with excluded limitations regarding significant changes in their construction.

According to § 64 (1)–(2) WHG, the one which is allowed to discharge over 750 $m^3$ of wastewater daily is obliged to designate one or several representatives for water protection. The competent body can decide to implement deviations from this rule. Such representatives' tasks include, in particular, giving advice to a water user in matters related to observing the proper regulations and effectiveness of procedures associated with proper wastewater discharge and wastewater protection.

Specific requirements regarding the chemical composition of wastewater, the content of harmful substances, and similar technical norms are set out in the regulation concerning the requirements concerning wastewater discharge to waters Verordnung über Anforderungen an das Einleiten von Abwasser in Gewässer (Abwasserverordnung – AbwV; see reference under 6.1). It is noteworthy that the detail level for the composition and amount of harmful substances in wastewater required 110 pages of text and tables in appendices for this regulation.

Chapter 6 WHG mentions specific forms of water law violations that are punishable by a fine. These include: hindering natural water flow, discharging wastewater to rivers or sewage installations without the relevant permit, and unauthorised construction or alteration of sewage installations, § 103 (1) n. 9 and 10 WHG. This is punishable by a fine of up to € 50.000, § 103 (2) WHG.

**Legal provisions concerning wastewater management in individual federal states**

As already explained in the introduction, water resources legislation is subject to the concurrent legislative competence of the federal government in accordance with Article 72 (1) and (3), Article 74 (1) n. 32 GG. Accordingly, the federal states are authorised to supplement federal regulations of the WHG and the Federal Wastewater Ordinance with their own regulations or to adopt deviating regulations. The individual federal states have made use of this option.

One of the examples is Wassergesetz für Baden-Württemberg (WG BW) of 2013. Wastewater removal is a task of a commune (but this does not apply to federal or state waterways or ones that go beyond a town) or decentralised discharges. If wastewater management is not a burden for a commune, then wastewater disposal is the responsibility of the entity which produces it, § 46 (1–2) WG BW. Communes can agree on a concept of wastewater management with a competent water authority as an internal planning instrument (*internes Planunginstrument*). Such a concept provides a review of the status of wastewater removal and describes its planned development. This concept should describe the way of rainwater disposal, what sites will be allotted for central wastewater storage tanks, when they will be used, and what sites will be used for decentralised ones. This concept should be publicised in a locally accepted way (§ 47 WG BW). Construction and use of wastewater installations should comply with the WHG provisions described earlier.

An entity which uses public wastewater installations should keep a register of the wastewater amount and its impact on the installation, § 4 (1) WG BW. Like WHG, private wastewater installations require that proper staff should be hired to operate them.

Wastewater collection in Saxony, regulated at SächsWG, is also a responsibility of communes but also of public law units. The level of detail of the provisions concerning wastewater is different here than in Baden-Württemberg, as they also apply, e.g., to sludge from mines with no outflow or wastewater from construction sites. Hence, responsibilities associated with wastewater management are described in a broader subjective aspect. In general, a water authority permit is required to perform actions associated with wastewater removal. An entity conducting wastewater management should develop a concept for wastewater removal, which must be approved by the competent water authority. There is the noteworthy § 54 (1–3) SächsWG: an entity which removes wastewater is obliged to perform an analysis of its status at least once a month, at its own expense. The laboratory is indicated in the wastewater discharge permit. The Saxon law contains detailed provisions concerning the types of wastewater installations. They take into consideration their use for business purposes in mining or as a result of their construction.

As mentioned, some federal states have issued further legal regulations, in particular ordinances for the area of wastewater disposal, on the basis of their state water laws:

**Baden-Württemberg:** Verordnung des Umweltministeriums über das Einleiten of Abwasser in öffentliche Abwasseranlagen (*Indirekteinleiterverordnung* – IndVO) of 19 April 1999, lately amended by Gesetzes vom 3 December 2013.

**Bavaria:** Verordnung zum Umsetzung der IVU-Richtlinie bei Abwasser (*Bayerische* IVU-*Abwasser-Verordnung* – IVUAbwV) of 12 December 2001.

**Berlin:** Verordnung über das Einleiten von Abwasser in öffentliche Abwasseranlagen (*Indirekteinleiteveordnung* – IndV) of 1 April 2005, lately amended by Article 43 der Verordnung of 1 September 2020.

**Brandenburg:** Verordnung über die Behandlung von kommunalen Abwasser im Land Brandenburg (*Brandeburgische Kommunalabwasserverordnung* – BbgKAbwV) of 18 February 1998, lately amended by Article 2 Absatz 26 des Gesetzes of 25 January 2016.

**Bremen:** Verordnung über die Behandlung von kommunalem Abwasser (KomAbwV) of 23 April 1997.

**Hamburg:** Hamburgisches Abwassergesetz (HmbAbwG) of 24 June 2001, lately amended by Article 5 des Gesetzes of 23 January 2018.

**Hesse:** Verordnung über das Einleiten von Grundwasser und Abwasser in öffentliche Abwasseranlagen (*Indirekteinleitunverordnung* – IndV) of 20 June 2023.

**Mecklenburg-Western Pomerania:** Verordnung über die Behandlung von kommunalen Abwasser (*Kommunalabwasserverordnung* – KabwVO M-V) of 15 December 1997, lately amended by Article 17 Abs. 15 des Gesetzes of 27 May 2016.

**Rhineland-Palatinate:** Landesverordnung über die Beseitigung von kommunalem Abwasser (KomAbwVO) of 27 November 1997, lately amended by § 136 des Gesetzes of 14 July 2015.

**North Rhine-Westphalia:** Verordnung zur Selbstüberwachung von Abwasseranlagen – Selbstüberwachungsverordnung Abwasser (SüwVO Abw) of 17 October 2013, lately amended by Article 2 des Gesetzes of 4 May 2021.

**Saarland:** Verordnung über die Eigenkontrolle von Abwasserbehandlungsanlagen (*Eigenkontrollverordnung* – EKVO) of 18 February 1994, lately amended by Article 175 des Gesetzes of 8 December 2021.

**Saxony:** Eigenkontrollverordnung vom 7 October 1994, lately amended by Artikel 9 des Gesetzes of 12 July 2013.

**Saxony-Anhalt:** Verordnung über die Selbstüberwachung von Abwasseranlagen und Abwassereinleitungen (*Selbstüberwachungsverordnung* – SÜVO) of 5 August 2021.

**Thuringia:** Thüringer Verordnung über das Einleiten oder Einbringen von Abwasser nach § 49 Abs. 2 des Thüringer Wassergesetzes in öffentliche Abwasseranlagen (*Thüringer Indirekteinleiverordnung* – ThürIndEVO) of 8 March 2000, lately amended by Article 5 des Gesetzes of 28 May 2019.

**Summary**

Provisions of the WFD were implemented into the German law in Gesetz zur Ordnung des Wasserhaushalts of 2009 (*Wasserhaushaltsgesetz* – WHG). The requirement of sustainable water use from Article 1b WFD is reflected in § 1 WHG, according to which the law pursues the purpose of 'sustainable water management'. Nevertheless, water law in Germany is federally characterised. The reason for this is the provisions in the Constitution (*Grundgesetz*) on legislative competence. The power to legally regulate the water balance is transferred to the federal government as a concurrent competence in accordance with Articles 72, 74 (1) n. 32 GG. If the federal government has made use of this authorisation, Article 72 (3) n. 5 GG authorises the individual federal states to make deviating provisions. However, according to explicit constitutional provisions, this does not apply to substance and plant-related regulations. Nevertheless, the federal government, as the holder of concurrent legislative power, retains the right to concurrent intervention. According to Article 72 (3), 3 GG, the later law takes precedence over the earlier one.

The WHG regulates integrated water management in river basin districts and thus implements the corresponding requirements of the WFD, and, in principle, the use of a body of water requires a permit or authorisation from the competent authority. With regard to water management in §§ 27–31 WHG the basic management concept of the WFD for surface waters can be found. In connection with this, the provisions on water management planning – namely, the programme of measures (§ 82) and the management plan (§ 83) – are to be seen. On the basis of the WHG 2009, the individual federal states have finalised the updating of their water laws.

Sections 54–61 of the WHG contain the basic provisions on wastewater disposal, some of which are specified in the Federal Wastewater Ordinance and the provisions of the Water Act of the federal states and the ordinances of the federal states based on these. The overarching principle is that wastewater should be discharged in such a way that 'the public good is not impaired'. The public good is the objective of the WHG and therefore the decisive criterion for all wastewater disposal measures. The overarching principle is that wastewater should be discharged in such a way that 'the public good is not impaired'. The public good is the objective of the WHG and therefore the decisive criterion for all wastewater disposal measures. Wastewater facilities must comply with the state of the art, and, in principle, the discharge of wastewater into bodies of water and wastewater facilities requires authorisation.

Enforcement of water law is the responsibility of the federal states. As the administrative structure is regulated differently in the federal states, the question of which authority is specifically responsible cannot be answered in a standardised manner. Nevertheless, there are parallels in the administrative organisation. In each federal state, the responsible ministry or the corresponding authority is the supreme authority for water management. The majority of the tasks are assigned to the lower state authorities, i.e., the districts, independent cities, towns, and municipalities belonging to districts.

## References

Berendes, K. (2018). *WHG. Wasserhaushaltsgesetz. Kurzkommentar*. 2. Edition. Erich Schmidt Verlag GmbH & Co. KG.

Bogumil, J., and W. Jann (2020). *Verwaltung und Verwaltungswissenschaft in Deutschland*. 3. Edition. Springer VS.

Czychowski, M., and M. Reinhardt (2019). *Wasserhaushaltsgesetz. Kommentar*. 12. Edition C.H. Beck OHG.

Dallhammer, W.-D., B. Dammert, and K. Faßbender (2018). *Sächsisches Wassergesetz. Kommentar für die Praxis*, 1. Edition. Kohlhammer,

Degenhart, C. (2023). *Staatsrecht I Staatsorganisationsrecht*. 39. Edition. C.F. Müller,

Drost, U. (2023). *Das neue Wasserrecht. Kommentare mit Vorschriftensammlung*. Loose-leaf collection as of 25th supplementary edition. Boorberg.

Hömig, D., and H.A. Wolff (2022). *Grundgesetz für die Bundesrepublik Deutschland*. 13. Edition. Nomos.

Jarass, H.D., B. Pieroth, and M. Kment (2024). *Grundgesetz für die Bundesrepublik Deutschland. Kommentar*. 18. Auflage, C.H. Beck OHG.

Wellmann, S.R., P. Queitsch, and K.-D. Fröhlich (2019). *Wasserhaushaltsgesetz. Kommentar*. 2. Edition. Wiesbaden.

# 7 Legal protection of surface waters in Poland

*Elżbieta Zębek*

## Introduction

The Water Framework Directive (WFD) provisions were implemented into the Water Law of 20 July 2017 (WL) in Poland. However, like in other European Union (EU) countries, there were some issues concerning its implementation, arising mainly from the lack of sufficient tools for environmental and economic assessment, gaps in individual river basin areas or individual categories of waters, insufficient funds, and deficiencies in monitoring programmes (Maciejewski & Walczykiewicz, 2006; Pawlaczyk, 2012; Zębek, 2022). The water law regulates water management in accordance with the sustainable development principle, in particular, shaping and protecting water resources, water use, and management (Article 1). Its provisions apply to inland surface waters, which are divided into standing and flowing waters (Article 21). According to the classification provided in Article 22 of the WL, inland flowing waters include (a) natural watercourses (rivers) and springs that give rise to such watercourses, (b) lakes and other natural water reservoirs with constant or periodical natural inflow or outflow of surface waters, (c) artificial water reservoirs situated on flowing waters, and (d) canals. Inland standing waters include lakes and other natural water reservoirs not directly connected in a natural way to inland surface flowing waters (Article 23; Ćwiek et al., 2017, 88).

Polish law uses two terms: 'managing waters' and 'administering water resources'. Managing waters is a broader term, and it denotes the comprehensive treatment of water resources, both in their qualitative and quantitative aspect, by their use and protection. According to Article 9 of the WL, water administration should be performed while observing the principle of rational and comprehensive treatment of surface and groundwater, considering their quantity and quality. Water administration takes into consideration the common interest principle and requires the cooperation of public administration, water users, and representatives of local communities insofar as it allows for the maximum social benefits to be achieved. Water management is performed in line with the public interest while preventing any avoidable deterioration of the ecological function of waters and the status of land ecosystems dependent on water. Water resource administration, according to

DOI: 10.4324/9781003586913-8

Article 10 of the WL, aims at satisfying the needs of a community and of the economy and at protecting waters and the environment linked to these resources, in particular in terms of (a) supplying water in a sufficient amount and of sufficient quality; (b) protection against drought and flood; (c) protection of water resources against pollution and improper or excessive exploitation; (d) maintaining or improving the status of aquatic ecosystems and those dependent on waters; (e) supplying water for agriculture and industry; (f) creating conditions for using waters in power generation, transport, and fisheries; (g) satisfying needs related to tourism, sport, and recreation.

## Legal basis of surface water management and protection

Instruments of water resource management in the Polish law include (1) planning in water management, (2) water law permits, (3) water service fees and other charges, (4) water management control, and (5) water management information system (Article 11 of the WL). Water resource administration is effected while taking into account a division of the territory of the state into drainage basins, water regions, and river basins (Article 12 of the WL). Therefore, the water law establishes instruments for water resource administration and protection, as well as obligations of public administration and control bodies, in particular of Polish Waters, which is the major institution administering water resources. The aim of these instruments of water resource administration is to perform tasks that meet the needs of the population and the economy and to protect waters and the environment linked to these resources. Planning in water management is to facilitate programming and coordinating actions aimed at (a) achieving and maintaining a good status of waters and ecosystems dependent on waters, as well as protection, improvement, and prevention of further deterioration of aquatic and land ecosystems and wetlands; (b) improvement of the water resource status; (c) reducing the number of substances and energy which may have an adverse impact on water; and (d) achieving environmental goals. The water-law permits performing the function of rationing and protecting the use and protection of water resources. Pursuant to Article 388, water-law permits may have three forms, i.e., a permit, a notification, and/or water-law assessment. The catalogue of actions for which a water-law permit is required is contained in Articles 34 and 389. They include activities as part of the special use of waters, e.g., wastewater discharge to waters or to soil in an amount exceeding $5m^3/d$ (Article 34), as well as water services, discharging industrial wastewater containing particularly hazardous substances to wastewater systems, construction of water engineering structures (Article 389). The aim of a water-law assessment is to analyse and assess the impact of a planned investment or action on the possibility of achieving environmental goals. The obligation to seek an assessment precedes filing an application for a water-law permit. Pursuant to Article 425 sect. 1 of the WL, a water-law assessment is required for investments or using water services, introducing substances that inhibit algae development to inland surface waters or constructing water engineering structures. A detailed list of investments and actions that require a water-law assessment is provided in the Regulation of 27 August 2019 on the types of investments and actions that require a water-law assessment. Control of water management is another legal instrument of surface water administration. Its scope is defined in Article 334 of the WL, and it includes water use and protection of water resources, observing conditions set out in decisions issued under the law (water-law permits), construction of water engineering structures, and maintenance of water and water engineering structures. The water management information system is used for gathering information on water resources, listed in Article 329 of the WL, i.e., concerning the borders of a drainage basin, the amount and quality of groundwater resources, the size of surface water intake, and many others. One must emphasise the importance of entering and disposal of data in the water management information system for proper water resource administration (Ćwiek, 2017; Pikor, 2019; Rotko, 2019; Szuwalski, 2019; Korzeniowski, 2021; Zębek et al., 2016; Zębek, 2020; Zębek 2021).

The aim of water protection is to achieve the environmental goals for surface water bodies, groundwater bodies and protected areas, as well as to improve water quality and biological relations in an aquatic environment and in wetlands (Article 51). Pursuant to Article 56 of the WL, the environmental goal for surface water bodies not identified as artificial or heavily modified, including rivers, is to protect and improve their ecological and chemical status so as to achieve at least good ecological status and good chemical status and to prevent deterioration of their ecological and chemical status. Classification of surface waters is based on the Regulation of 25 June 2021 on the classification of the ecological status, ecological potential and chemical status, and methods of classification of the status of surface water bodies and environmental quality standards for priority substances. The regulation lists physicochemical, biological, and hydromorphological elements for individual categories of water bodies, taking into account various types of surface waters, elements of quality for classification of surface water body ecological status in natural water courses (e.g., rivers), lakes, and other natural water reservoirs, method of classification of these elements, assessment, and presentation of the water status, frequency of water quality classification (§ 1). According to the regulation, the assessment is based on a 5-point water classification (classes I–V) based on assessment of the chemical and ecological water status in the following scale: class I – very good, class II – good, class III – moderate, class IV – poor, and class V – bad (Giakoumis & Voulvoulis, 2018; Zębek, 2017; Zębek & Napiórkowska-Krzebietke, 2019).

## Water authority – responsibilities and competences

The structure, responsibilities, and competences of the water authority in Poland are set out mainly in Articles 14 and 240 of the WL. The authorities competent in water management matters include (1) the minister for water management, (2) the minister for inland navigation, (3) chairman of Polish Waters, (4) director of a regional water management board of Polish Waters, (5) director of a river basin board of Polish Waters, (6) head of water supervision at Polish Waters, (7) director of maritime authority, (8) voivode, (9) county head, and (10) commune head, mayor (Article 14 of the WL).

Water management tasks are performed by entities of three kinds: water administration bodies, water authorities, and State Water Holding Polish Waters (*Wody Polskie*) in three ways because each entity functions and operates differently. The State Water Holding Polish Waters is an exception, as it combines the competences of the imperium sphere and performs the tasks in the dominium sphere. A comparison of the competences and tasks of water administration bodies and water authorities shows that the former perform tasks of an imperious nature. On the other hand, water authorities perform public tasks with no imperious elements (Rakoczy, 2018). This provides grounds for stating that the State Water Holding Polish Waters, with the bodies making up its structure, is the most important entity administering water resources. The legislator assigned specific competences concerning water management to each of these organs in Article 240 of the WL. The chairman of Polish Waters is the central organ of the entity, whereas the others have the status of local organs.

The minister for water management is the supreme organ of government administration with competences in water management. These include

1. submitting to the Polish Sejm, every two years, not later than on 31 August, information on water management concerning (a) the status of water resources owned by the state and their use, (b) implementation of water management plans in drainage basins, (c) international cooperation on border waters and execution of agreements in this regard, (d) performance of tasks regarding the maintenance of waters and the other property of the State Treasury associated with water management, (e) investments, and (f) implementation of flood risk management plans and plans for counteracting the effects of drought;
2. shaping the state's water policy, taking into consideration the inland navigation development;
3. coordination of the performance of public tasks in water management; this, in particular, concerns providing guidelines to and instructing the chairman of Polish Waters regarding methods of task performance and accepting proposals of the chairman of Polish Waters regarding methods of performing investment tasks;
4. supervising the activities of Polish Waters;
5. fulfilling information and reporting obligations towards the European Commission within the scope specified in the law (Articles 354–356 of the WL);
6. performing duties arising from international agreements on water management, to which Poland is a party; and
7. supervising the activities of state services (Article 360 of the WL).

The director of the regional board of Polish Waters is competent to make such decisions as the following:

- Order restoration of the status of non-polluting waters in the areas of special flood risk (Article 77 sect. 11)
- Agreements regarding building up and development of the land situated in the area particularly threatened with flooding (Article 166 sect. 5)
- Ordering a water impounding facility to lower the water level or to empty the reservoir in the event of a flash flood warning (Article 182 sect. 1)
- Agreements concerning funding or co-funding of drainage facilities at the cost of the State Treasury, with the participation of community resources (Article 199 sect. 4)
- Establishing and adjusting the amount of a drainage fee (Articles 201 and 202)
- Issuing a decision on discontinuation of operation of a facility or its part which might result in a hazard to human, animal life or health, or to the environment, or to a legally protected interest of third persons, until the negligence is remedied (Article 343 sect. 2)
- Banning the use of water for a facility operating without a required water-law permit (Article 343 sect. 6)

The director of a regional water management board may also issue water-law permits in the following cases: (a) special use of waters, use of water services, construction of water engineering structures, or operation of water installations or engineering structures for enterprises with a potentially considerable environmental impact – group I; (b) for the construction of anti-flood structures; (c) for transfer of water and for the construction of the necessary water engineering structures; (d) for discharging substances inhibiting algae growth to surface waters; (e) for actions associated with rehabilitation of surface waters; (f) for excavating rock, gravel, sand, or other materials from surface waters, and for cutting down plants in the surface waters or on the banks; (g) for discharging industrial wastewater containing substances of particular environmental harmfulness to wastewater systems; and (h) water-law assessments in the aforementioned scope (Article 397 sect. 3 of the WL).

The director of a drainage basin board of Polish Waters is competent in matters of decisions concerning:

1 Allowing for introducing rainfall water from melting snow or ice overflow or wastewater from stormwater drainage overflow of general urban wastewater system if this does not collide with the environmental objectives for waters or quality requirements for waters (Article 80)
2 Distributing of costs of maintenance of structures or walls that are not water engineering structures, making up the water bank (Article 226 sect. 4)

3 Determining the amount of a fixed fee for water services (Article 271 sect. 7)
4 Establishing the amount of surface water intake or the amount of wastewater discharged to waters or to soil without a required water-law permit (Article 281 sect. 7)

Moreover, the director may issue water-law permits in cases other than those mentioned in Article 397 sect. 3 of the WL, e.g., for surface water intake or for wastewater discharge into waters in an amount not exceeding a total of 5 $m^3$ per day (Article 33 sect. 4), water use for business activities other than agriculture (Article 34 sect. 13), or construction of water engineering structures (Article 389 sect. 6).

The head of water supervision at Polish Waters is competent to receive a water-law application, for example, for (a) construction of a pier of up to 3 m in width and total length of up to 25 m, which is the total length of its individual elements; (b) construction of ponds that are not filled with rainfall waters or groundwater as part of water services, with the area of up to 5000 $m^2$ and the depth not exceeding 3 m from the natural surface of the land; and (c) alteration of a ditch which involves the construction of a culvert or a different closed section along the length of not more than 10 m (Article 394 sect. 2). The director may also set the requirements for water management regarding the conditions of construction of water engineering structures or other actions that require notification, as long as it is justified by the need to protect waters or the environment (Article 424 sect. 1).

The voivode's responsibilities include (a) developing voivodship plans for crisis management in cooperation with regional water management boards (Article 240 (3)), (b) adopting a water maintenance plan as an act of bye-law at a request of Polish Waters (Article 327 sect. 4), and (c) establishing the amount of financial compensation in a decision to compensate for damage other than damage caused by flooding of land, which took place, in particular, in connection with executing a water-law permit or filing a water-law application (Article 469 sect. 1).

The county head has a right to apply for selling land under inland standing waters owned by the State Treasury (Article 216 sect. 4), issues opinions about appointing and dismissing the head of water supervision, approves a statute of a water company, supervises and controls the activities of a water company (Article 446–467).

If the owner, by altering the state of water on his land, has caused damage to neighbouring land, the commune head or mayor may order the land owner to restore the water to its previous state or to make devices to prevent such damage (Article 234 sec. 3). In addition, the authority shall set the amount of the fee for the reduction of natural land retention as a result of carrying out on a property with an area of more than 3,500 $m^2$ works or construction facilities permanently connected to the land having the effect of reducing such retention by excluding more than 70% of the area of the property from the biologically active area in areas not covered by open or closed drainage systems (Article 272 sec. 22).

According to Article 239 sect. 3 of the WL, Polish Waters comprises the following units:

1 Water management board with a registered seat in Warsaw
2 Regional water management boards with offices in Białystok, Bydgoszcz, Gdańsk, Gliwice, Krakowie, Lublin, Poznań, Rzeszów, Szczecin, Warsaw, and Wrocław
3 River basin boards
4 Water supervisory boards

These tasks are performed by Polish Waters under the provisions of the Regulation of 28 December 2017 on granting the statute to the State Water Holding Polish Waters in three areas: (1) protection against flood and drought, (2) water services, and (3) aquatic environment management. Activities in the first area include matters related to flood and drought, i.e., planning, project preparation, and execution of investments, as well as maintaining and operating hydrotechnical objects. They also include matters related to water supply for agriculture and monitoring the hydrological situation and crisis situations. The tasks in the water service area include matters related to water users, administrative proceedings for water law consents, including receiving water-law notifications and issuing water-law permits, charging fees for water services, and controlling water management. Activities in the area of aquatic environment administration include matters related to the implementation of EU directives, e.g., WFD, including preparing draft water management plans in drainage basins and issues related to protected sites, e.g., Nature 2000, as well as the information system and water management monitoring (State Water Holding Polish Waters, 2023; Zębek & Napiórkowska-Krzebietke, 2023).

## Legal regulations concerning wastewater management

The legislation sets out the rules for discharging liquid waste (wastewater) in order to protect surface waters. A definition of wastewater in accordance with Article 16 sect 16 of the WL is very broad, and it includes (a) water used up for household or business purposes, (b) leachate from landfills and mining waste disposal sites, where hazardous mining waste and mining waste other than hazardous or neutral is stored, sites for storage, retrieval, or neutralisation of waste, used brine, medicinal, and thermal waters; (c) water from cooling down circulation in power or thermal power plants (CHP); and (d) water from dehydration of mines. Protection of surface waters against pollution as per Article 54 sect. 1 of the WL should primarily involve the reduction of emission to waters from spot pollution sources with the use of permissible emission levels, resulting from the best available techniques, which should not be exceeded within a specified time period. It must be pointed out that the permissible emission levels denote a mass, concentration, or emission level for energy or substances specified in the regulations issued on the basis of Article 99 (1) of the WL. The Regulation 12 July 2019 on substances particularly harmful to the aquatic environment and the conditions which must be fulfilled when wastewater is discharged to water or to soil and when rainwater or water from melting snow and ice is channelled away to water or to water engineering structures lays down the permissible emission levels for substances particularly harmful to the aquatic environment, causing water pollution which should be eliminated (list I) and substances particularly harmful to the aquatic environment, causing water pollution which should be reduced (list II). The regulation sets out the conditions which must be fulfilled when discharging wastewater to water or to soil, in particular household wastewater, urban wastewater, and industrial wastewater, including (a) the highest permissible levels of pollutants and the conditions which must be fulfilled for agricultural wastewater use, as well as the place, method, and minimum frequency of wastewater sample collection, reference methodologies of analysis, and method for assessment whether the wastewater meets the conditions; (b) the highest permissible levels of pollutants in wastewater treatment plants for household wastewater and urban wastewater; and (c) the conditions to be fulfilled when rainwater or water from melting snow and ice is channelled away to water or to water engineering structures, including the highest permissible levels of pollutants and the place, the method and the minimum frequency of wastewater sample collection, reference methodologies of analysis, and the method for assessment whether rainwater or water from melting snow and ice fulfils the conditions. According to Article 75 of the WL, wastewater discharge is forbidden to natural water courses and to canals with water-inflowing lakes if the time of wastewater inflow to the lake is shorter than 24 hours. Wastewater discharged to water or to the soil as part of ordinary water use or water services should be treated to the extent required by the law, and they cannot contain priority substances or cause changes in their natural, characteristic biocenosis, changes of the natural cloudiness, colour, or smell, as well as the formation of deposits or foam (Article 78 of the WL). Entities discharging wastewater to waters or to soil have to provide water protection against pollution, in particular by construction and operation of devices used for such protection, and – where reasonable – reuse of the treated wastewater. Moreover, the choice of a place and method of use or removal of waste should minimise the adverse environmental impact (Article 83 of the WL; Zębek, 2012). There are no clear procedures for handling emergency situations which involve uncontrolled inflow of wastewater containing substances particularly harmful to surface waters.

Industrial activities are executed as part of special water use, requiring a water-law permit, e.g. water services, for introducing substances particularly harmful to the aquatic environment with industrial wastewater to wastewater systems (Article 34 of the WL), construction of water engineering structures preceded by obtaining a water-law assessment (Article 425). According to Article 270 of the WL, a fee for water services, i.e., for wastewater discharge to waters or to soil, consists of a fixed fee and a variable fee dependent on the amount and quality of wastewater discharged under a water-law permit. The fee for wastewater discharge to waters or to soil depends on the type of substance contained in wastewater and its amount, as well as on the type of wastewater. The amount of the fixed fee provided for in Article 271 for wastewater discharge to waters or to soil is established by Polish Waters (the river basin board director), which is then communicated to entities obliged to incur the payments for water services as annual information, also containing the method for calculating the payment. The fixed fee for wastewater discharge to waters or to soil is calculated by multiplying the unit fee, the time expressed in days, and the maximum amount of wastewater discharged to waters or to soil, specified in the water-law permit or in an integrated permit and expressed in $m^3/s$. The upper unit fee for water services in the case of a fixed fee for wastewater discharge to water or to soil is PLN 500 per day per 1 $m^3/s$ for the maximum amount of wastewater discharged to waters or to the soil, set out in the water-law permit or an integrated permit (Article 274).

Community participation in water protection is regulated in the Act on Access to Information on the Environment and Its Protection, the Community Participation in Its Protection and on Environmental Impact Assessments of 3 October 2008. Community participation is required in the case of group I enterprises, i.e., those that can always have a considerable environmental impact, which is part of the proceedings for an environmental impact assessment for a planned enterprise (Article 3 sect. 1, item 8) before a decision is issued on environmental factors. Pursuant to Article 72 (1), an environmental permit is issued before a water-law permit is obtained for the construction of water engineering structures, e.g., outlets of wastewater devices used for wastewater discharge to water, soil, or water engineering structures and outlets used for water discharge to water, soil or water engineering structures (Article 16 sect. 65f of the WL). A list of enterprises which require an environmental impact assessment is provided in the Regulation of 10 September 2019 on enterprises with a potential considerable environmental impact. Pursuant to § 3 (1), industrial wastewater treatment plants, except installations which do not discharge to waters or to engineering structures of wastewater containing substances of particular harmfulness to the aquatic environment, are classified as enterprises with a potential considerable impact on the environment, i.e., of group II, not requiring an environmental impact assessment (EIA) or community participation. Community participation is provided for in the WL in the procedure of developing plans for counteracting the effects of drought (Article 153 sect. 3), plans for water management in a drainage basin (Article 318), plans for water maintenance that require a strategic EIA (Article 327), a procedure for a water-law permit, preceded by an environmental permit (Article 402), as well as for enterprises with a potential considerable environmental impact, for which a water-law assessment is replaced by an environmental decision and a decision preceded by an environmental impact reassessment (Article 428).

### Control and monitoring of surface water management

Pursuant to Article 335 of the WL, control of water management is performed mainly by Polish Waters with respect to using water and protecting water resources, observing the conditions set out in permits granted under the law, observing the conditions set out in integrated permits, constructing water engineering structures, maintaining waters and water engineering structures, observing obligations and restrictions imposed on landowners, and observing the conditions valid in protective zones and areas. Control of water management over Polish Waters in this respect is exercised by the minister for water management. The authorities exercising the control and the minister competent for water management cooperate in performing the control activities with other control bodies, public administration bodies, and community organisations, as well as with Border Guards (Articles 342 and 343 of the WL). Based on the findings of a control, the controlling authority or the minister for water management can (a) issue post-control recommendations, (b) request that the competent body should initiate administrative proceedings as provided for in the law, (c) request joining ongoing proceedings as a party, and (d) impose a fine for petty offences specified in the law (Article 341). Pursuant to Article 343, the director of a regional water management board of Polish Waters calls for a facility to remedy within a specific time period the negligence in water management which might result in a hazard to human, animal life or health, or to the environment, or to a legally protected interest of third persons. If the negligence is not remedied within this time period, this organ issues a decision on discontinuation of the facility or its part activities until the negligence is remedied. If it is found during the control that the facility operates without the required water-law permit, the organ can ban it from water use. This decision is enforceable immediately. Pursuant to Article 476 sect 1 of the WL, a deed which consists of using waters, i.e., construction of water engineering structures or performing other actions that require a water-law permit (e.g., wastewater discharge into surface waters) without a required water-law permit or in violation of the conditions set out in the water-law permit, is punishable by detention, restriction of liberty, or a fine. Moreover, pursuant to Article 478 sect 5 of the WL, a deed consisting of wastewater discharge into waters or soil when such wastewater contains a substance of particular harmfulness to the environment, thus violating the conditions laid down in regulations on industrial wastewater discharge, is punishable by a fine. Judgement is carried out in accordance with the rules laid down in the Act of 24 August 2001 – Code of Procedure for Petty Offences.

The aim of surface water monitoring pursuant to Article 349 sect. 1 of the WL is to gather information on the status of such waters and on the status of protected areas for the purpose of planning in water management and assessment of progress in attaining environmental goals. Surface waters are examined, and their status is assessed as part of the state environment monitoring. A competent organ of the Environmental Protection Inspection performs an examination of surface waters with respect to biological, physicochemical, and chemical agents, including priority substances. Based on the findings of tests and observations, this organ makes an assessment of the status of surface water bodies in the territory of a voivodship, taking into account transitional and coastal waters, as well as in the drainage basin. The supervising entity (regional environmental protection director) for the areas assigned for habitat or species protection, designated in the Nature Conservation Act of 16 April 2004, for which water status maintenance or improvement is an important factor in their protection, makes an assessment of these areas based on examinations of the objects of protection dependent on the water status and the monitoring findings. The Environmental Protection Inspection operates under the Act of 20 July 1991 on Environmental Protection Inspection. According to Article 3 of the act quoted above, organs of the Environmental Protection Inspection include the Chief Environmental Protection Inspector as a central organ of government administration supervised by the minister competent for climate and the voivodship environmental protection inspector as an organ of government administration in a voivodship. When it comes to emergency situations, the Environmental Protection Inspection cooperates in removing a major failure with organs competent to remove it and supervises the removal of the effects of such a failure (Article 30). However, this applies to the supervision and control of industrial plants, and there is no reference to emergency situations which involve surface water pollution with hazardous substances. The Environmental Protection Inspection monitors the environmental status according to the rules set out in the Regulation of 13 July 2021 on forms and methods of monitoring surface water bodies and groundwater bodies. According to § 4 of the regulation, there are the following types of monitoring of surface water bodies: (1) diagnostic, based on planning documentation and it concerns the planned status assessment for surface water bodies; (2) operational, based on planning documentation or diagnostic monitoring and concerns a status assessment for surface water bodies regarded as threatened with failure to attain specific environmental goals; (3) investigational, to explain the cause of failure to attain environmental goals; and (4) for protected areas.

Polish legislation introduced some legal and organisational solutions for water monitoring and information system after the environmental disaster in the Oder River in 2022. The causes of the event included a set of weather, hydrological, and environmental factors, as well as those resulting from anthropogenic activities, i.e., high temperature, low water levels and slow water flow, high insolation, and very high values, including chloride, sulfate, and sodium ions, which could indicate the impact of mining activities. This has contributed to the massive development in the form of a water bloom of the so-called golden alga – *Prymnesium parvum* – a species characteristic of saline environments in the seas (e.g., Norwegian Sea, Baltic Sea) and in industrial areas. This species produced fish-killing toxins called prymnesiums, causing damage to their gills, which was the cause of a massive fish die-off in this river of 249 tons. As part of the remediation of the Oder River, recommendations have been developed, including the establishment of a system of continuous measurement of water quality for selected parameters associated with blooms, as well as periodic monitoring of the algae *Prymnesium parvum* and the fauna (including fish, mollusks) for environmental contamination (dioxins, heavy metals, pesticides, toxic compounds; IOŚ-PIB, 2022; Zębek & Napiórkowska-Krzebietke, 2023). In this regard, the General Inspector for Environmental Protection (GIEP) has developed an appropriate procedure for monitoring the presence of golden algae in the Oder River. The procedure takes into account 3 degrees of danger (I–III), indicating, respectively, medium, high, and very high risk of a bloom caused by the presence of *Prymnesium parvum.* In addition, an information transfer system has been developed with three stages:

1 Field measurements and sampling using reference methods and laboratory tests – CLB branch (Central Laboratory of GIEP) acting on behalf of PIEP/GIEP (Provincial Inspector of Environmental Protection/General Inspector of Environmental Protection)
2 Hazard level analysis – CLB (GIEP National Reference Laboratory for Surface Water Quality), information will be provided to DEM (GIEP Department of Environmental Monitoring) and then to CAT (GIEP Control and Audit Team) on exceeding alarm levels
3 External information – 'golden alga' report – CAT to notify the assigned alert/threat of bloom levels to the CMC MCE (Crisis Management Center of the Ministry of Climate and Environment), NWMB SWMC PL (National Water Management Board of the State Water Management Company Polish Water), GDEP (General Director of Environmental Protection), CMI (Central Mining Institute) and the territorially competent PCMC (Provincial Crisis Management Center) and PIEP (GIEP, 2023)

The Oder River Revitalization Law was enacted in 2023, which regulates detailed rules and conditions for the preparation of investments aimed at improving the environmental conditions of the Oder River in terms of the quantity and quality of water in the river, as well as the aquatic environment, and increasing the water retention capacity of the river's catchment area. Also included are measures to determine the extent of anthropogenic impact on the Oder River and rules for implementing measures to restore ichthyofauna in the river (Article 1). However, the law focuses more on regulating the riverbed than on restoring ichthyofauna. Article 6 indicates a system for providing information on the occurrence of a hazard or event that may cause a change in water quality at a water intake used to supply water for human consumption to the public or in a water intake recharge area. The system includes sanitary inspectors, environmental inspectors, and the water authority. Article 7 contains provisions requiring the Polish Water Authority to act on the Oder River Basin Management Plan. The catalogue of these activities first includes (1) an analysis of the feasibility of reconstructing damming structures to ensure biological and morphological continuity and meet environmental objectives on rivers, (2) an assessment of the impact of transverse structures on the biological continuity and environmental objectives of surface water bodies, (3) control of the operation of fish migration devices to determine their proper operation and allow for adequate fish migration, and (4) monitoring of the effectiveness of existing fish migration devices to verify the proper operation of these devices. If negative impacts of structures on biological continuity and environmental objectives are identified - analyses of the possibility of implementing measures to ensure biological continuity and meet environmental objectives. As a result of the enactment of the Oder River Revitalization Law, the catalogue of reasons for revoking or limiting a water permit contained in Article 415(9) of the WL has been expanded in the event that further use of

> water under the conditions established in the water permit creates a condition that endangers human life or health, or a condition that threatens to cause serious damage to the environment, in particular due to existing changes in the aquatic environment.

The introduction of this provision may contribute to faster action, i.e., by the authority responsible for water permits (Gruszecki, 2024). The Water Inspectorate was also established as part of water control under Article 251a of the WL (effective 1 July 2024). The Water Inspectorate is headed by the National Water Inspector with the assistance of regional water inspectors. The President of the Polish Water Authority serves as the National Water Inspector. Among the tasks of the National Water Inspectorate is to determine the risk of potential events that may cause adverse effects on waters or fish stocks in waters in order to eliminate these effects (Article 251 c).

## Summary

The Water Law in Poland specifies in detail the environmental goals to match those laid down in the WFD with respect to the protection of the waters, including the introduction of liquid waste (wastewater), and creates a complex system of legal instruments. The most important instruments include planning in water management, water-law permits, water management control, and information systems, as well as a system for fees for water services (including industrial wastewater discharge to surface waters) and surface water monitoring. However, the Water Law in Poland contains a multitude of regulations, which are often formulated too generally, as a result of which the same responsibilities and competences are assigned to different organs and units of Polish Waters. This applies, for example, to issuing decisions on discontinuation of the facility or its part activities until the negligence is remedied and water management control tasks without specifying the scope of such controls. The regulations do not take into account sufficiently crisis situations, which only include floods, but not, e.g., intensive pollution of water reservoirs by an expected wastewater discharge, which can be regarded as an extraordinary situation. There were no special procedures that had been developed for such situations (Zębek & Napiórkowska-Krzebietke, 2023). However, after the ecological disaster in the Oder River, legislative and organisational measures were taken to restore the ichthyofauna. As part of these measures, procedures were developed by the GIEP for monitoring the state of the waters, giving a scale of 3 degrees of threat based on the chemical state of the waters and the abundance of *Prymnezium parvum* (golden alga), as well as the creation of a comprehensive procedure for sampling, laboratory analysis, assessment of the state of threat, and the sequence of transmission of information to various authorities and institutions. This system will allow early assessment of the threat of golden alga blooms and the smooth flow of information about the threat. In addition, the catalogue of prerequisites for withdrawal or restriction of a water permit was expanded, which applies to entities that cause the further use of water under the conditions set forth in the water permit to create a condition that endangers the life or health of people or a condition that endangers the occurrence of serious damage to the environment, in particular due to the occurrence of changes in the aquatic environment. The introduction of this provision may contribute to faster action, i.e., by the authority responsible for water permits. A Water Inspectorate has also been established to determine the risk of potential events that may cause adverse effects on waters or fish stocks in waters in order to eliminate these effects.

There is another matter of a system of punishments for wastewater discharge into waters or soil when such wastewater contains substances of particular harmfulness to the environment, thus violating the conditions concerning industrial wastewater discharge. Such a deed is punishable by a fine, i.e., it is a petty offence. Therefore, the maximum fine pursuant to art. 1 § 1 of the Code of Minor Offences of 1971 amounts to PLN 5,000, whereas the administrative fines provided for in the Waste Act of 14 December 2012 for handling waste contrary to the provisions of this act range from PLN 1,000 to PLN 1,000,000, which seems to be a more severe punishment for a business entity than a fine. The last issue under analysis is the classification of industrial wastewater treatment plants, which discharge to waters or to facilities wastewater containing substances of particular harmfulness to the aquatic environment, as enterprises with a potentially considerable impact on the environment, i.e., of group II, not requiring an EIA or community participation. However, the EIA procedure should be obligatory for such enterprises.

## References


Ćwiek, P. (2017). Ocena wodnoprawna – nowe obowiązki inwestora. LEX/el.

Ćwiek, P., A. Cybulska, L. Osuch-Chacińska, E. Piętowska, M. Stachowska, and M. Wyszogrodzki,(2017). *Nowe Prawo wodne.*Omówienie. Pytania i odpowiedzi. Tekst ustawy. Wolters Kluwer.

Giakoumis, T., and N. Voulvoulis (2018). Progress with monitoring and assessment in the WFD implementation in five European River Basins: significant differences but similar problems. European Journal of Environmental Sciences 8 (1):44–50. https://doi.org/10.14712/23361964.2018.7

GIEP. (2023). Procedura monitorowania interwencyjnego Prymnesium parvum złotej algi, 3. Edition, 7 November 2023. https://www.google.com/search?client=firefox-b-d&q=Procedura+monitorowania+interwencyjnego+Prymnesium+parvum+%E2%80%9Ez% C5%82otej+algi%E2%80%9D

Gruszecki, K. (2024). *Prawo wodne.* Gospodarowanie wodami. Komentarz. LEX WKP.

IOŚ-PIB. (2022). Wstępny raport zespołu ws. sytuacji na rzece Odrze, 30 September 2022, edited by A. Kolada. https://ios.edu.pl/aktualnosci/wstepny-raport-zespolu-ds-sytuacji-na-rzece-odrze

Korzeniowski, P. (2021). *Zgoda wodnoprawna.* LEX.

Maciejewski, M., and T. Walczykiewicz (2006). Dotychczasowe doświadczenia związane z wdrażaniem Ramowej Dyrektywy Wodnej. Infrastruktura i Ekologia Terenów Wiejskich 4 (1):63–76.

Pawlaczyk, P. (2012). Perspektywy osiągnięcia celów Ramowej Dyrektywy Wodnej w Polsce. Przegląd Przyrodniczy 23 (3):52–68.

Pikor, D. (2019). Pozwolenie wodnoprawne – procedura uzyskania. LEX/el.

Rakoczy, B. (2018). *Prawo wodne.* Praktyczny przewodnik. LEX WKP.

Rotko, J. (2019). Ocena wodnoprawna w świetle przepisów ustawy z 20.07.2017 r. – Prawo wodne. *Business Law Journal* 72 (10):2–9. https://doi.org/10.33226/0137-5490.2019.10.1

State Water Holding Polish Waters (2023). Responsibilities of Polish Waters. https://www.wody.gov.pl/o-wodach-polskich/zadania-wod-polskich

Szuwalski, P. (2019). Prawo wodne. Komentarz do wybranych przepisów. LEX/el.

Zębek, E. (2012). Ocena oddziaływania na środowisko a ochrona wód przed zanieczyszczeniem. *Studia Prawnoustrojowe* 18:173–183.

Zębek, E. (2017). Międzynarodowe i krajowe podstawy prawne i bioindykatory (glony) oceny stanu jakości wód powierzchniowych. In Odpowiedzialność za środowisko w ujęciu normatywnym, edited by E. Zębek and M. Hejbudzki, 119–138. UWM Press.

Zębek, E. (2020). Water-law permission as an administrative and legal instrument for the management and protection of water resources. Acta Scientiarum Polonorum Administratio Locorum 19 (2):119–130. https://doi.org/10.31648/aspal.4866

Zębek, E. (2021). Ocena wodnoprawna jako nowy instrument administracyjno-prawny ochrony wód powierzchniowych. *Gubernaculum et Administratio* 1 (23): 191–210. https://doi.org/10.16926/gea.2021.01.12

Zębek, E. (2022). Legal solutions of lake monitoring systems in Poland in compliance with the Water Framework Directive. *Review of European and Comparative Law* 48 (2):173–201. https://doi.org/10.31743/recl.13419

Zębek, E., and A. Napiórkowska-Krzebietke (2019). Rozwój przepisów prawnych w zakresie bioindykacji środowiskowej a stan jakości wód jeziorowych. *Studia Prawnoustrojowe* 43:375–393. https://doi.org/10.31648/sp.4616

Zębek, E., and A. Napiórkowska-Krzebietke (2023). Działania prawne i naprawcze wobec szkód w środowisku wodnym. Studium przypadku katastrofy ekologicznej na rzece Odrze. Zeszyty Prawnicze Biura Analiz Sejmowych Kancelarii 3 (79): 48–72. https://doi.org/10.31268/ZPBAS.2023.52

Zębek, E., M. Szwejkowska, and M. Raczkowski (2016). Pozwolenie wodnoprawne jako instrument reglamentacyjno-ochronny w użytkowaniu zasobów wodnych w działalności gospodarczej w kontekście sprawiedliwości ekologicznej. In *Sprawiedliwość ekologiczna w prawie i praktyce*, edited by T. Bojar-Fijałkowski, 355–363. Fundacja Rozwoju Uniwersytetu Gdańskiego.

# Conclusion

*Elżbieta Zębek and Dominik Zidek*

The Water Framework Directive (WFD) is regarded as a very modern approach to water management, taking into account the basic principles of environmental law such as the principles of appropriate development, the principle of subsidiarity, the polluter-pays principle, and the principle of social participation. As water protection legislation continues to develop in European law, there is also a significant change in the assessment of their quality. This legislation comprises not only standards that aim to regulate pollution and other activities harmful to the environment but also to prevent, reduce or counteract risks to the environment, including the aquatic environment. Thus, European Union (EU) legislation deals with water quality standards, pollution control, and water resource management principles.

Despite the problems of implementing the provisions of the WFD in the legislation of EU member states, they can be considered to have taken all measures to achieve good surface water status by establishing more or less effective legal instruments for the management and protection of surface waters. Important legal instruments are planning, the water permit system, the control and monitoring system, and crisis management procedures. However, we should not forget about the important role of water authorities in achieving environmental objectives through measures established in national law.

The analysis of solutions and legal instruments for the management and protection of surface waters in selected EU countries, i.e., the Czech Republic, Germany, France, Lithuania, Spain, and Poland, shows their great diversity, which is probably conditioned by political, geographical, and ecological conditions. These countries have varying levels of renewable surface water resources, from the lowest in Poland and the Czech Republic to the highest in Lithuania (at the EU average). Therefore, it can be assumed that most of the countries discussed are experiencing problems with available water resources. This process may intensify in the future in the form of water deficits on a larger scale. The water legislation established in these countries, influenced by the WFD, is characterised by a certain specificity and originality, and the legal instruments used also differ in their effectiveness, as evidenced by the recorded different levels of achievement of good status of surface waters (rivers and lakes). Therefore, it is justified for the authorities

DOI: 10.4324/9781003586913-9

of these countries to strive to create effective legal instruments in the field of protection of these waters, both in terms of quantity and quality.

When comparing the solutions and legal instruments used for the management and protection of surface waters in the countries analysed, certain similarities and differences can be identified. The identical solutions include, among others, the catalogue of water resources management instruments, i.e., planning, water permit system, monitoring, and control. Another issue is the responsibility for monitoring and emergency situations: the Environmental Inspectorate in Poland and the Czech Republic and the Environmental Protection Agency in Lithuania. There are also similarities in the structure of the water authority as a superior body at the ministerial level in the Czech Republic, Poland, France, Spain, and Lithuania. In addition, water management is controlled by the higher water supervisory authorities in the Czech Republic, France, and Poland (the highest authority for Polish Waters). However, the fundamental difference between the countries analysed is the structure of water authorities, the management of river basins and watercourses, and the charges for water services. In Poland, for example, it can be considered centralised within the Polish Waters Institution, whose bodies have most of the powers to the detriment of local government administrative bodies. In Spain, there is also a separate body – the National Water Management Council – as the central level of the river basin organisation, which has the greatest powers in the management of water resources. In other countries, the authority over water is shared between the national government, the autonomous communities (Spain), and the federal states (Germany). The competences of the water authorities in the field of the instruments of surface water management (water-law permits, planning, information system, control, and monitoring) are also different. A greater role for local governments in water resource management is evident in the Czech Republic and France. In the Czech Republic, separate bodies have also been set up for the management of river basins and the management of watercourses. The bodies and responsibilities for protecting surface water from pollution also vary considerably. Other bodies have the power to suspend the operation of an industrial plant, and not all the countries analysed (e.g., Lithuania) have developed procedures for dealing with emergency situations of water pollution with harmful substances. These procedures have been implemented in an exemplary manner, e.g., in the Czech Republic and most recently in 2023 after the ecological disaster in the Odra River – also introduced in Poland. Some countries also have specialised units that assist in the control and monitoring of surface waters, such as the Water Police in Spain and the Water Inspectorate in Poland. In addition, there are irrigation communities in Spain, which are justified by the need for irrigation of land due to the specific climate. In France, local governments and environmental associations play an important advisory role. Although other countries also require public consultation when assessing the impact on aquatic ecosystems of major investments that also serve water protection purposes, these procedures are carried out only at the basic and necessary level required by law.

Based on the previous analysis, it can be concluded that despite some issues concerning the implementation of the WFD by the EU countries analysed in this study, many regulations concerning water law were passed, which could be a model for other countries and contribute to the improvement of surface water quality. This comparative legal analysis has served as the basis for the formulation of a number of recommendations. The first thing that can be pointed out is the uniformity of the water-law regulations in one legal act, which comprehensively regulates the matter, as is the case in Spain, but with the leading role of the law of environmental protection, which should regulate only the matters common to the whole system. The rules and all the aspects of water management and surface water protection should be regulated in one comprehensive legal act. Such as the Code of Environment in France (*Fr. Code de l'environnement*). In addition, a good solution seems to be simplifying the system of water authority, as is the case in the water law in the Czech Republic; assigning clear competence will contribute to more effective performance of tasks by the organs and units competent in water management. To improve the effectiveness of water quality management and protection, member states should streamline and consolidate responsibilities across their various ministries and administrative bodies. The current fragmentation leads to inefficiencies and delays in response, as seen in the Bečva River accident in the Czech Republic. Establishing a centralised coordinating body or enhancing the authority of existing ones could ensure more coherent and rapid action. Additionally, integrating all relevant regulations and directives into a more unified legislative framework would reduce overlaps and gaps, facilitating better compliance and enforcement. Moreover, a change of functions of the minister for water management and deconcentration, and maybe decentralisation of the administrative competence, while expanding the authority to create the state policy, like in Spain. The legal status of the State Water Management Council has also changed, and it no longer plays the role of an advisory body to the minister for water management as part of the activities of the country's national consultative and participatory forum. In addition, it would be necessary to increase the role of local government, with which the institutional aspect is linked, as in the French regulations. A specific feature of the French regulations is the important role of a prefect who – to put it simply – performs a function similar to that of a voivodship governor, which is known in Polish law. This should be regarded as a very useful and right solution, as it allows for the central authorities to influence the water management system. Similarly, the role of local authorities as organs participating in water and wastewater management should be increased. The experience of the German legal system proves the high effectiveness of such distribution of the water authority's competence. In general, it is the local organs that implement local water management plans, which are publicised. Applications can be submitted to them by the entities concerned.

Among the positive solutions, it is also worth mentioning an important part of the German water management system: its openness to changing circumstances, which is a consequence of technological development and knowledge of water management. This manifests itself in leaving considerable freedom to public administration organs (by discretionary actions) concerning the taking of the necessary actions. With relative freedom to decide on water and wastewater management, including supervision and control, administration organs can act effectively, taking into account the general federal guidelines regarding water management needs. In addition, there would also have to be an increase in the role of case law, including *Conseil d'Etat's* power to issue decree laws, which allows for creating rather general regulations, which are then made more specific by case law. It is also necessary to recommend a change in the nature of the water cooperatives and flood bank associations towards real water-irrigation cooperatives, as is the case in Spain, but using the previous experience of Polish land cooperatives and agricultural self-government (agricultural chambers) or social and professional farmer organisations. In the EU water-law system, there should be detailed regulation of the status and authority of organisations associated with environmental protection. They play an important role in French law, provided that they have been operating for at least three years, and they have a status of approved environmental protection associations (fr. *associations agréées de protection de l'environnement*).

An important issue in water law is the protection of water against pollution. It is worth noting that distribution of the sources of water law in the German federal legal system can seem to be somehow unclear. Nevertheless, it is the individual states that take into account the local specificity of the water management needs, including the method for wastewater removal and the distribution of the legal responsibility associated with it. This aspect cannot be simply taken into account in other legal systems, but it does not exclude the possibility of increasing the role of local authorities in wastewater discharge. In this regard, it is necessary to impose additional obligations towards the entities that discharge industrial wastewater to water reservoirs, as is the case in the water law of the Czech Republic, especially as regards the requirements concerning the obtaining of a water-law permit and installations. In addition, the extension of the catalogue of grounds for withdrawing or limiting a water permit seems reasonable, as in Poland, which applies to entities that cause further use of water under the conditions set forth in the water permit to create a condition that endangers the life or health of people or a condition that endangers the occurrence of serious damage to the environment, in particular due to the occurrence of changes in the aquatic environment. The introduction of this provision may contribute to faster action, i.e., by the authority responsible for water permits. In addition, the establishment of a Water Inspectorate to determine the risk of potential events that may cause adverse effects on waters or fish stocks in waters in order to eliminate these effects. In this regard, it is necessary to implement special rules for handling particularly harmful substances (preventive and corrective actions) which enter surface waters and pose a threat to the aquatic environment, as well as the procedures for emergency situations in water reservoirs, as is the case in the Czech Republic. The process of liquidation of the effects should involve, to a greater extent, organs of public administration of a lower level (county and commune) and a suitable system for financing such actions should be established. Addressing specific pollution issues, particularly phosphorus contamination, requires stringent pollution control measures. Member states should enhance their regulatory framework to enforce stricter emission norms for pollutants and implement effective reduction programmes for surface water pollution. This includes revisiting and potentially tightening permits for industrial discharges and ensuring that water treatment technologies align with the best available practices. Continuous monitoring and assessment, coupled with adaptive management strategies, can help in achieving and maintaining good chemical and ecological status of surface waters. The introduction of appropriate procedures for dealing with emergencies specific to a particular water body (may also apply to other waters), as in Poland after the Oder environmental disaster for monitoring the state of waters, should be recommended. This procedure gives a scale of 3 degrees of threat based on the chemical state of the water and the abundance of *Prymnezium parvum* (golden algae), as well as the creation of a comprehensive procedure for sampling, laboratory analysis, assessment of the state of threat, and the sequence of transmission of information to various authorities and institutions. This system will allow early assessment of the threat of golden alga blooms and the smooth flow of information about the threat.

One of the most important issues in water law is the introduction of the administrative responsibility of an entrepreneur for violating the regulations concerning the functioning of wastewater installations and following the guidelines contained in the water use permit along with the material responsibility of an entity using waters for the damage done in the German water law. The severity of the financial penalty (a fine of up to 50,000 euros) is not proof of its effectiveness, but it serves the coherences of the system of administration in matters related to wastewater removal. There should be sanctions for illegal (not complying with the requirements) wastewater discharge to surface waters as a financial penalty in the Czech water law (in a specified range) depending on the concentration of substances in excess of the conditions laid down in the permit for wastewater discharge to waters, the range of water pollution, and the extent of protection of an affected area (protected areas) and the cause of the illegal water discharge. There is a need for stronger enforcement mechanisms and accountability, especially concerning environmental crimes. For example, Czechia should bolster its criminal liability framework for environmental offences to ensure that perpetrators, including businesses, face significant consequences for violations. This can involve increasing penalties, improving prosecution processes, and ensuring that regulatory bodies have the necessary resources and authority to enforce compliance effectively. Additionally, public awareness and participation in environmental protection efforts can be increased to foster a culture of responsibility and proactive engagement in safeguarding water resources.

The aforementioned recommendations of legal solutions implemented and functioning in the field of management and protection of inland waters in the analysed EU countries may provide an ace to change the overall approach in the area of both quantitative and qualitative water protection. This is especially very important in the face of climate change and increasing industrial pressure. The recommendations presented can also serve as a model for application in other EU countries facing similar problems in the field of water management and protection.

# Appendix 1

## List of legal acts

*European Union*

UN, Resolution No. A/RES/70/1 adopted by the General Assembly on 25 September 2015 Transforming our world: the 2030 Agenda for Sustainable Development.

European Parliament Resolution of 8 September 2015 on the follow-up to the European Citizens' Initiative Right2Water (2014/2239(INI)) (OJ C 316/99 P8_TA(2015)0294, 22.9.2017).

Resolution 48/13 of the United Nations Human Rights Council of 8 October 2021 on the human right to a clean, healthy and sustainable environment.

Resolution of the European Parliament of 5 October 2022 on access to water as a human right (2021/2187(INI)).

Resolution No. 2020/2613(RSP) on the implementation of the EU water legislation of 17 December 2020, P9_TA(2020)0377.

Convention on the Protection and Use of Transboundary Watercourses and International Lakes, signed in Helsinki on March 17, 1992 (LJ of 2003, No. 78, item 702).

International Watercourses adopted by the General Assembly of the United Nations on 21 May 1997.

Treaty on the Functioning of the European Union (consolidated version) (OJ C 326, 26.10.2012, 47–390).

Council Directive 91/271/EEC of 21 May 1991 concerning urban wastewater treatment (OJ L 135, 30.5.1991, 40–52).

Council Directive 98/83/EC of 3 November 1998 on the quality of water intended for human consumption (OJ L 330, 5.12.1998, 32–54).

Directive 2000/60/EC of the European Parliament and of the Council of 23 October 2000 establishing a framework for Community action in the field of water policy (OJ L 327, 22.12.2000, 1 as amended).

Directive 2004/35/CE of the European Parliament and the Council of 21 April 2004 on the environmental liability with regard to the prevention and remedying of environmental damage (OJ L 143, 30.4.2004, 56–75).
Directive 2007/2/EC of the European Parliament and of the Council of 14 March 2007 establishing an Infrastructure for Spatial Information in the European Community (INSPIRE) (OJ L 108, 25.4.2007, 1–14)
Directive 2008/105/EC of the European Parliament and of the Council of 16 December 2008 on environmental quality standards in the field of water policy, amending and subsequently repealing Council Directives 82/176/EEC, 83/513/EEC, 84/156/EEC, 84/491/EEC, 86/280/EEC and amending Directive 2000/60/EC of the European Parliament and of the Council (OJ L 348, 24.12.2008, 84–97).
Directive 2009/90/EC of 31 July 2009 laying down, pursuant to Directive 2000/60/EC of the European Parliament and of the Council, technical specifications for chemical analysis and monitoring of water status (OJ L 201, 1.8.2009, 36–38).
Directive (EU) 2020/2184 of the European Parliament and of the Council of 16 December 2020 on the quality of water intended for human consumption (OJ L 435/123, 23.12.2020, 1–62).
Decision (EU) 2022/591 of the European Parliament and of the Council of 6 April 2022 on a General Union Environment Action Programme to 2030, PE/83/2021/REV/1 (OJ L 114, 12.4.2022, 22–36).
Regulation (EC) 1907/2006 of the European Parliament and of the Council of 18 December 2006 concerning the Registration, Evaluation, Authorisation and Restriction of Chemicals (REACH), establishing a European Chemicals Agency, amending Directive 1999/45/EC and repealing Council Regulation (EEC) 793/93 and Commission Regulation (EC) 1488/94 as well as Council Directive 76/769/EEC and Commission Directives 91/155/EEC, 93/67/EEC, 93/105/EC and 2000/21/EC (OJ L 393, 30.12.2006, 1 as amended).

## *Czechia*

Ústavní zákon č. 1/1993 Sb., Ústava České republiky.
Usnesení předsednictva České národní rady o vyhlášení LISTINY ZÁKLADNÍCH PRÁVA SVOBOD jako součásti ústavního pořádku České republiky.
Zákon č. 2/1969 Sb., o zřízení ministerstev a jiných ústředních orgánů státní správy České republiky.
Zákon č. 282/1991 Sb., o České inspekci životního prostředí a její působnosti v ochraně lesa.
Zákon č. 114/1992 Sb., o ochraně přírody a krajiny.
Zákon č. 114/1995 Sb., o vnitrozemské plavbě.
Zákon č. 305/2000 Sb., o povodích.
Zákon č. 254/2001 Sb., o vodách a o změně některých zákonů (vodní zákon).

Zákon č. 100/2001 Sb., o posuzování vlivů na životní prostředí a o změně některých souvisejících zákonů (zákon o posuzování vlivů na životní prostředí).
Zákon č. 274/2001 Sb., o vodovodech a kanalizacích pro veřejnou spotřebu a o změně některých zákonů (zákon o vodovodech a kanalizacích).
Zákon č. 500/2004 Sb., správní řád.
Zákon č. 167/2008 Sb., o předcházení ekologické újmě a o její nápravě a o změně některých zákonů.
Zákon č. 40/2009 Sb., trestní zákoník.
Zákon č. 418/2011 Sb., o trestní odpovědnosti právnických osob a řízení proti nim.
Zákon č. 255/2012 Sb., o kontrole (kontrolní řád).
Zákon č. 250/2016 Sb., o odpovědnosti za přestupky a řízení o nich.
Zákon č. 283/2021 Sb., stavební zákon.
Zákon č. 148/2023 Sb., o jednotném environmentálním stanovisku.
Zákon č. 182/2024 Sb., kterým se mění zákon č. 254/2001 Sb., o vodách a o změně některých zákonů (vodní zákon), ve znění pozdějších předpisů, zákon č. 114/1992, o ochraně přírody a krajiny, ve znění pozdějších předpisů, a zákon č. 465/2023 Sb., kterým se mění zákon č. 416/2009 Sb., o urychlení výstavby dopravní, vodní a energetické infrastruktury a infrastruktury elektronických komunikací (liniový zákon), ve znění pozdějších předpisů, a další související zákony.
Nařízení vlády č. 401/2015 Sb., o ukazatelích a hodnotách přípustného znečištění povrchových vod a odpadních vod, náležitostech povolení k vypouštění odpadních vod do vod povrchových a do kanalizací a o citlivých oblastech.
Vyhláška č. 104/1997 Sb., kterou se provádí zákon o pozemních komunikacích.
Vyhláška č. 450/2005 Sb., o náležitostech nakládání se závadnými látkami a náležitostech havarijního plánu, způsobu a rozsahu hlášení havárií, jejich zneškodňování a odstraňování jejich škodlivých následků.
Vyhláška č. 49/2011 Sb., o vymezení útvarů povrchových vod.
Vyhláška č. 98/2011 Sb., o způsobu hodnocení stavu útvarů povrchových vod, způsobu hodnocení ekologického potenciálu silně ovlivněných a umělých útvarů povrchových vod a náležitostech programů zjišťování a hodnocení stavu povrchových vod.
Vyhláška č. 50/2023 Sb., o plánech povodí a plánech pro zvládání povodňových rizik.

*France*

Arrêté n° 0059 du 5 mars 2015 précisant les critères et méthodes d'évaluation de la teneur en nitrates des eaux et de caractérisation de l'enrichissement de l'eau en composés azotés susceptibles de provoquer une eutrophisation et les modalités de désignation et de délimitation des zones vulnérables définies aux articles R. 211-75, R. 211-76 et R. 211-77 du code de l'environnement.

Décret n°92-1042 du 24 septembre 1992 portant application de l'article 5 de la loi n° 92-3 du 3 janvier 1992 sur l'eau et relatif aux schémas d'aménagement et de gestion des eaux, décret modifié par le décret n°2005-1329 du 21 octobre 2005.

Décret n°2005-475 du 16 mai 2005 relatif aux schémas directeurs d'aménagement et de gestion des eaux.

Décret n°2005-613 du 27 mai 2005 relatif à l'évaluation des incidences de certains plans et programmes sur l'environnement.

Décret n°2005-636 du 30 mai 2005 relatif à l'organisation de l'administration dans le domaine de l'eau et aux missions du préfet coordonnateur de basin.

Décret n°2005-1329 du 21 octobre 2005 pris en application des articles L.212-3 à L.212-7 du Code de l'environnement et modifiant le décret n°92-1042 du 24 septembre 1992.

Décret n°2006-880 du 17 juillet 2006 relatif aux procédures d'autorisation et de déclaration prévues les articles L. 214-1 à L. 214-3 du Code de l'environnement pour la protection de l'eau et des milieux aquatiques.

Décret n°2007-135 du 30 janvier 2007 précisant les critères de définition et de délimitation des zones humides figurant à l'article L. 211-1 du Code de l'environnement.

Décret n°2007-882 du 14 mai 2007 relatif à certaines zones soumises à contraintes environnementales et modifiant le code rural.

Décret n°2007-1213 du 10 août 2007 relatif aux schémas d'aménagement et de gestion des eaux et modifiant le code de l'environnement.

Décret n°2007-1557 du 2 novembre 2007 relatif aux installations nucléaires de base et au contrôle, en matière de sûreté nucléaire, du transport de substances radioactives.

Décret n°2007-1760 du 14 décembre 2007 portant dispositions relatives aux régimes d'autorisation et de déclaration au titre de la gestion et de la protection de l'eau et des milieux aquatiques, aux obligations imposées à certains ouvrages situés sur les cours d'eau, à l'entretien et à la restauration des milieux aquatiques et modifiant le code de l'environnement.

Décret n°2007-1872 du 26 décembre 2007 relatif à l'affectation du débit artificiel des cours d'eau à certains usages.

Décret n° 2011-2018 du 29 décembre 2011 portant réforme de l'enquête publique relative aux opérations susceptibles d'affecter l'environnement.

Décret n°2011-2021 du 29 décembre 2011 déterminant la liste des projets, plans et programmes devant faire l'objet d'une communication au public par voie électronique.

Décret n°2012-616 du 2 mai 2012 relatif à l'évaluation de certains plans et documents ayant une incidence sur l'environnement.

Décret n°2015-1038 du 20 août 2015 relatif aux établissements publics territoriaux de bassin et aux établissements publics d'aménagement et de gestion de l'eau.

Décret n°2016-1071 du 3 août 2016 relatif au schéma régional d'aménagement, de développement durable et d'égalité des territoires.

Décret n° 2018-847 du 4 octobre 2018 relatif aux schémas directeurs d'aménagement et de gestion des eaux et schémas d'aménagement et de gestion des eaux.

La loi n°2004-338 du 21 avril 2004 portant transposition de la directive 2000/60/CE du Parlement européen et du Conseil du 23 octobre 2000 établissant un cadre pour une politique communautaire dans le domaine de l'eau est parue au J.O n° 95 du 22 avril 2004 page 7327.

La loi de 2 mars 1982 nr 82-213 – Loi relative aux droits et libertes des communes, des departements et des regions.

Act No. 64-1245 of 16 December 1964 on water regime and distribution and on combating its pollution (*Fr. Loi n°64-1245 du 16 décembre 1964 relative au régime et à la répartition des eaux et à la lutte contre leur pollution*).

Act No. 2000-108 of 10 February 2000 on modernisation and development of public electric service (*Fr. Loi n°2000-108 du 10 février 2000 relative à la modernisation et au développement du service public de l'électricité*).

Act No. 92-3 of 3 January 1992 on water (*Fr. Loi n°92-3 du 3 janvier 1992 sur l'eau*).

Act No. 2004-338 of 21 April 2004 transposing Directive 2000/60/EC of 23 October 2000 establishing a framework for Community action in the field of water policy (*Fr. Loi n°2004-338 du 21 avril 2004 portant transposition de la directive 2000/60/CE du 23 octobre 2000 établissant un cadre pour une politique communautaire dans le domaine de l'eau*).

Act No. 2006-1772 of 30 December 2006 on waters and aquatic environment (*Fr. Loi n°2006-1772 du 30 décembre 2006 sur l'eau et les milieux aquatiques*).

Programme act No. 2009-967 of 3 August 2009 on holding the environmental forum Grenelle (*Fr. Loi n°2009-967 du 3 août 2009 de programmation relative à la mise en œuvre du Grenelle de l'environnement*).

Act No. 2010-788 of 12 July 2010 on the state's commitment to protecting the environment (*Fr. Loi n°2010-788 du 12 juillet 2010 portant engagement national pour l'environnement*).

Act No. 2014-58 of 27 January 2014 on modernisation of territorial public actions and affirmation of a metropolis (*Fr. Loi n°2014-58 du 27 janvier 2014 de modernisation de l'action publique territoriale et d'affirmation des métropoles – MAPTAM*).

Act No. 2014-110 of 6 February 2014 on better regulating the use of phytosanitary products on the national territory (*Fr. LOI n° 2014-110 du 6 février 2014 visant à mieux encadrer l'utilisation des produits phytosanitaires sur le territoire national*).

Act No. 2015-991 of 7 August 2015 on new territorial organisation of the Republic (*Fr. Loi n°2015-991 du 7 août 2015 portant nouvelle organisation territoriale de la République – NOTRe*).

Act No. 2016-1087 of 8 August 2016 on restoring biodiversity, nature and landscapes (*Fr. Loi n°2016-1087 du 8 août 2016 pour la reconquête de la biodiversité, de la nature et des paysages*).

***Spain***

Constitución Española (BOE-A-1978-31229).

Decreto 356/2010, de 3 de agosto, por el que se regula la autorización ambiental unificada, se establece el régimen de organización y funcionamiento del registro de autorizaciones de actuaciones sometidas a los instrumentos de prevención y control ambiental, de las actividades potencialmente contaminadoras de la atmósfera y de las instalaciones que emiten compuestos orgánicos volátiles, y se modifica el contenido del Anexo I de la Ley 7/2007, de 9 de julio, de Gestión Integrada de la Calidad Ambiental (Boletín número 157 de 11/08/2010).

Decreto 239/2011, de 12 de julio, por el que se regula la calidad del medio ambiente atmosférico y se crea el Registro de Sistemas de Evaluación de la Calidad del Aire en Andalucía (Boletín número 152 de 04/08/2011).

Decreto-ley 3/2023, de 25 de abril, por el que se aprueban medidas adicionales para paliar los efectos producidos por la situación de excepcional sequía a los usuarios de las demarcaciones hidrográficas intracomunitarias de Andalucía y se adoptan medidas urgentes, administrativas y fiscales, de apoyo al sector agrario (Boletín Extraordinario número 13 de 27/04/2023).

Ley de Aguas, de 13 de junio de 1879. Gaceta del 19 de Junio de 1879.

Ley 10/2001, de 5 de julio, del Plan Hidrológico Nacional (BOE-A-2001-13042).

Ley 62/2003, de 30 de diciembre, de medidas fiscales, administrativas y del orden social (BOE-A-2003-23936).

Ley 11/2005, de 22 de junio, por la que se modifica la ley 10/2001 del Plan Hidrológico Nacional (BOE-A-2005-10622).

Ley Orgánica 2/2007, de 19 de marzo, de reforma del Estatuto de Autonomía para Andalucía (BOE-A-2007-5825).

Ley 7/2007, de 9 de julio, de Gestión Integrada de la Calidad Ambiental (BOE-A-2007-15158).

Ley 39/2015, de 1 de octubre, del Procedimiento Administrativo Común de las Administraciones Públicas (BOE-A-2015-10565).

Ley 1/2008, de 27 de noviembre, de medidas tributarias y financieras de impulso a la actividad económica de Andalucía, y de agilización de procedimientos administrativos (BOE-A-2008-20752).

Ley 4/2010, de 8 de junio, de aguas de la Comunidad Autónoma de Andalucía (BOE-A-2010-11490).

Ley 9/2010, de 30 de julio, de Aguas para Andalucía (BOE-A-2010-13465).

Ley 40/2015 de Régimen Jurídico del Sector Público (BOE-A-2015-10566).

Real Decreto 849/1986, de 11 de abril, por el que se aprueba el Reglamento del Dominio Público Hidráulico, que desarrolla los títulos preliminar I, IV, V, VI y VII de la Ley 29/1985, de 2 de agosto, de Aguas (BOE-A-1986-10638).

Real Decreto-ley 11/1995, de 28 de diciembre, por el que se establecen las normas aplicables al tratamiento de las aguas residuales urbana (BOE-A-1995-27963).

Real decreto 509/1996, de 15 de marzo, de desarrollo del Real Decreto-ley 11/1995, de 28 de diciembre, por el cual se establecen las normas aplicables al tratamiento de las aguas residuales urbanas (BOE-A-1996-7159).
Real Decreto Legislativo 1/2001, de 20 de julio, por el que se aprueba el texto refundido de la Ley de Aguas (BOE-A-2001-14276).
Real Decreto 907/2007, de 6 de julio, por el que se aprueba el Reglamento de la Planificación Hidrológica (BOE-A-2007-13182).
Real Decreto 1620/2007, de 7 de diciembre, por el que se establece el régimen jurídico de la reutilización de las aguas depuradas (BOE-A-2007-21092).
Real Decreto 60/2011, de 21 de enero, sobre las normas de calidad ambiental en el ámbito de la política de aguas (BOE-A-2011-1139).
Real Decreto 817/2015, de 11 de septiembre, por el que se establecen los criterios de seguimiento y evaluación del estado de las aguas superficiales y las normas de calidad ambiental (BOE-A-2015-9806).
Real Decreto Legislativo 1/2016, de 16 de diciembre, por el que se aprueba el texto refundido de la Ley de prevención y control integrados de la contaminación (BOE-A-2016-12601).
Real Decreto 1159/2021, de 28 de diciembre, por el que se modifica el Real Decreto 907/2007, de 6 de julio, por el que se aprueba el Reglamento de la Planificación Hidrológica (BOE-A-2021-21664).
A list of judgements
Tribunal Constitucional, Pleno, Sentencia 227/1988 de 29 Nov. 1988, Rec. 824/1985.
Supreme Court, Third Chamber for Contentious-Administrative Proceedings, 5th Section, Judgement of 26 Oct. 2011, Rec. 2810/2008. ECLI: ES:TS:2011:7182.

***Lithuania***

Lietuvos Respublikos Aplinkos Apsaugos Įstatymas (Lietuvos aidas 1992, nr. 20-0; Wiesz. 1992, nie. 5-75, i. k. 0921010ISTA00I-2223).
Lietuvos Respublikos Planuojamos Ūkinės Veiklos Poveikio Aplinkai Vertinimo Įstatymas (1996 m. rugpjūčio 15 d. Nr. I-1495).
Lietuvos Respublikos Vandens Įstatymas (Lietuvos aidas 1997 m. spalio 21 d. Nr. VIII-474).
Lietuvos Respublikos Jūros Aplinkos Apsaugos Įstatymas (Lietuvos aidas 1997 m. lapkričio 13 d. Nr. VIII-512).
Lietuvos Respublikos Aplinkos Monitoringo Įstatymas (Lietuvos aidas 1997 listopada 20 r. Nr VIII-529).
Lietuvos Respublikos Atliekų Tvarkymo Įstatymas (Lietuvos aidas 1998 m. birželio 16 d. Nr. VIII-787).
Nutarimas Dėl Nacionalinio Vandenų Srities 2022–2027 Metų Plano Patvirtinimo 2022 m. gruodžio 21 d. Nr. 1292.
Ordinance Nr. D2007-12 of the Minister of the Environment of the Republic of Lithuania of 1 April 2010 "Concerning the approval of the methodology for determination of the status of surface water bodies".

*Germany*

*Federal level*

Grundgesetz der Bundesrepublik Deutschland (GG) of 23.05.1949 (BGBl I S. 1, lately amended by Art. 1 des Gesetzes zur Änderung des Grundgesetzes vom 19.12.2022, BGBl. I S. 2478).

Bundeswasserstraßengesetz (WaStrG) of 23.5.2007 (BGBl. S. 962; 2008 I S. 1980, lately amended by Art. 5 des Gesetzes of 22.12.2023, BGBl. I Nr. 409).

Oberflächengewässerverordnung (OGewV) vom 20.06.2016 (BGBl I, S. 1373, lately amended by Art. 2 Abs. 4 des Gesetzes of 9.12.2020, BGBl. I S. 2873).

Verordnung über Anforderungen an das Einleiten von Abwasser in Gewässer (Abwasserverordnung – AbwV) of 17.06 2004 (BGBl. I S. 1108, 2625, lately amended by Art. 1 der Verordnung of 20.01.2022, BGBl. I S. 87).

Verordnung über Anlagen zum Umgang mit wassergefährdenden Stoffen (AwSV) vom 18.04.2017 (BGBl. S. 905, lately amended by Verordnung of 19.06.2020, BGBl. I S. 1328).

Wasserhaushaltsgesetz (WHG) of 31.07.2009 (BGBl. I S. 2585, lately amended by Art. 7 of the act of 22.12.2023, BGBl. I Nr. 409).

Level of the individual federal states

**Baden-Württemberg:**

Wassergesetz für Baden-Württemberg (WG BW) of 3.12.2013 (GBl. S. 389, lately amended by Art. 9 der Gesetzes vom 7.2.2023, GBl. S. 26, 43).

Verordnung des Umweltministeriums über das Einleiten von Abwasser in öffentliche Abwasseranlagen (Indirekteinleiterverordnung – IndVO) of 19. April 1999 (lately amended by Gesetzes of 3.12.2013, GBl. S. 389, 441).

Landesverwaltungsgesetz (VwG BW) of 14.10 2008 (GBl. S. 313, 314, lately amended by Art. 10 des Gesetzes of 21.5.2019, GBl. S. 161, 185).

**Bavaria:**

Bayerisches Wassergesetz (BayWG) of 25.2.2010 (GVBl. S. 66, 130, BayRS 753-1-U, lately amended by § 1 des Gesetzes of 9.11.2021, GVBl. S. 608).

Verordnung zum Umsetzung der IVU-Richtlinie bei Abwasser (Bayerische IVU-Abwasser-Verordnung – IVUAbwV) of 12.12.2001 (GVBl. S. 1066, BayRS 753-1-20-U).

**Berlin:**

Wassergesetz (BWG) i.d.F. der Bekanntmachung of 17.6.2005 (GVBl. 2005, 357; 2006, 248; 2007, 48, lately amended by Art. 2 des Gesetzes of 25.9.2019, GVBl. S. 612).

Verordnung über das Einleiten von Abwasser in öffentliche Abwasseranlagen (Indirekteinleiteveordnung – IndV) of 1.4.2005 (GVBl. 2005, 224, lately amended by Art. 43 der Verordnung of 1.9.2020, GVBl. S. 683).

**Brandenburg:**

Brandenburgisches Wassergesetz (BbgWG) in the version published on 2.3.2012 (GVBl. I/12, Nr. 20, lately amended by Art. 1 des Gesetzes of 4.12.2017, GVBl. I/17 Nr. 28).

Verordnung über die Behandlung von kommunalen Abwasser im Land Brandenburg (Brandeburgische Kommunalabwasserverordnung – BbgKAbwV) of 18.2.1998 (GVBl. II/98, Nr. 7, lately amended by Art. 2 (26) des Gesetzes of 25.1.2016, GVBl. I/16, Nr. 5).

**Bremen:**

Bremisches Wassergesetz (BremWG) of 12.4.2011 (Brem.GBl. S. 262, lately amended by Art. 6 no. 5 des Gesetzes of 24.11.2020, Brem.GBl. S. 1486, 1581).

Verordnung über die Behandlung von kommunalen Abwasser (KomAbwV) of 23.4.1997 (Brem.GBl. S. 172).

**Hamburg:**

Hamburgisches Wassergesetz (HWaG) in the version published on 29.3.2005 (HmbGVBl. 2005 S. 97, lately amended by Art. 12 des Gesetzes of 4.12.2012, HmbGVBl. S. 510. 519).

**Hessia:**

Hessisches Wassergesetz (HWG) of 14.12.2010 (GVBl. I S. 548, lately amended by Art. 39 des Gesetzes of 28.6.2023, GVBl. S. 473, 475).

Verordnung über das Einleiten von Grundwasser und Abwasser in öffentliche Abwasseranlagen (Indirekteinleitunverordnung – IndV) of 20.6.2023 (GVBl. S. 484).

**Lower Saxony:**

Niedersächsisches Wassergesetz (NWG) vom 19.2.2010 (Nds.GVBl. S. 64, lately amended by § 2 WasserG-Bagatellgrenze-ÄndVO of 6.12.2023, Nds. GVBl. S. 339).

Verordnung über Zuständigkeiten auf dem Gebiet des Wasserrechts (ZustVO-Wasser) of 10.3.2022 (Nds. GVBl. S. 70), lately amended by Art. 6 des Gesetzes of 19.12.2010, Nds. GVBl. S. 631).

**Mecklenburg-Vorpommern:**

Wassergesetz des Landes Mecklenburg-Vorpommern (LWaG) vom 30.11.1992 (GVOBl. M-V S. 669, lately amended by Gesetz of 8.7.2021, GVOBl. M-V S. 866).

Verordnung über die Behandlung von kommunalen Abwasser (Kommunalabwasserverordnung – KabwVO M-V) of 15.12.1997 (GVOBl. M-V 1998, 25, lately amended by Art. 17 Abs. 15 des Gesetzes of 27.05.2016, GVOBl. M-V S. 431, 440).

**North Rhine-Westphalia:**

Wassergesetz für das Land Nordrhein-Westfalen (Landeswassergesetz – LWG) in the version of 8.7.2016 (GV.NRW. S. 559, lately amended by Art. 3 des Gesetzes of 17.12.2021, GV. NRW. S. 1470).

Verordnung zur Selbstüberwachung von Abwasseranlagen – Selbstüberwachungsverordnung Abwasser (SüwVO Abw) of 17.10.2013 (GV. NRW. S. 602, lately amended by Art. 2 des Gesetzes of 04.05.2021, GV. NRW. S. 560).

**Rhineland-Palatinate:**

Landeswassergesetz (LWG) vom 14.7.2015 (GVBl. S. 127, lately amended by Art. 2 des Gesetzes vom 8.4.2022, GVBl. S. 118).

**Saarland:**

Saarländisches Wassergesetz (SWG) i.d.F. of 30.7.2004 (Amtsbl. S. 1994, lately amended by Art. 173 des Gesetzes of 8.12.2021, Amtsbl. I S. 2639).

Verordnung über die Eigenkontrolle von Abwasserbehandlungsanlagen (Eigenkontrollverordnung – EKVO) of 18.2.1994 (Amtsbl. 638, lately amended by Art. 175 des Gesetzes of 8.12.2021, Amtsbl. I S. 2629).

**Saxony:**

Sächsisches Wassergesetz (SächsWG) of 12.7.2013 (SächsGVBl. S. 503, lately amended by Art. 12 des Gesetzes of 20.12.2022, SächsGVBl. S. 705).

Eigenkontrollverordnung vom 7.10.1994 (SächsGVBl. S. 1592, lately amended by Artikel 9 des Gesetzes of 12.07.2013, SächsGVBl. S. 503).

Sächsische Wasserzuständigkeitsverordnung – SächsWasserZuVO of 12.6.2014 (SächsGVBl. S. 363, lately amended by Art. 13 HaushaltsbegleitG 2023/20224 of 20.12.2022).

**Saxony-Anhalt:**

Wassergesetz für das Land Sachsen-Anhalt (WG LSA) of 16.3.2011 (GVBl. LSA S. 492, lately amended by Art. 21 des Gesetzes of 7.7.2020, GVBl. LSA S. 372, 374).

Verordnung über die Selbstüberwachung von Abwasseranlagen und Abwassereinleitungen (Selbstüberwachungsverordnung – SÜVO) of 5.8.2021 (GVBl. LSA S. 457).

**Schleswig-Holstein:**

Landeswassergesetz (LWG) of 13.11.2019 (GVOBl. S. 425, § 101 amended by Art. 3 Nr. 3 Gesetzes of 6.12.2022, GVOBl. S. 1002).

**Thuringia:**

Thüringer Wassergesetz (ThürWG) of 28.5.2019 (GVBl. S. 74, lately amended by Art. 17 des Gesetzes of 11.6.2020, GVBl. S. 277, 285).

Thüringer Verordnung über das Einleiten oder Einbringen von Abwasser nach § 49 Abs. 2 des Thüringer Wassergesetzes in öffentliche Abwasseranlagen (Thüringer Indirekteinleiverordnung – ThürIndEVO) of 8.3.2000 (GVBl. S. 94, lately amended by Art. 5 des Gesetzes of 28.5.2019, GVBl. S. 74, 121).

Bundesverwaltungsgericht, Beschluss vom 19.12.1997 – 8 B 234/97 at NVwZ 1998, 1080 – 1082; DVBl. 1998, 1222 – 1223).

*Poland*

Code of Petty Offences of 20 May 1971 – *Ustawa z dnia 20 maja 1971 r. Kodeks wykroczeń* (consolidated text *Journal of Laws* of 2022 item 2151, as amended).

Act of 20 July 1991 on Environmental Protection Inspection – *Ustawa z dnia 20 lipca 1991 r. o Inspekcji Ochrony Środowiska* (consolidated text *Journal of Laws* of 2023, item 824).

Act of 24 August 2001 – Code of Procedure for Petty Offences – *Ustawa z dnia 24 sierpnia 2001 r – Kodeks postępowania w sprawach o wykroczenia* (consolidated text *Journal of Laws* of 2022, item 1124).

Act of 13 April 2007 on preventing damage to the environment and on fixing it – *Ustawa z dnia 13 kwietnia 2007 r. o zapobieganiu szkodom w środowisku i ich naprawie* (consolidated text *Journal of Laws* of 2020, item 2187).

Act of 3 October 2008 on Access to Information on the Environment and Its Protection, the Community Participation in Its Protection and on Environmental Impact Assessments – *Ustawa z dnia 3 października 2008 r. o udostępnianiu informacji o środowisku i jego ochronie, udziale społeczeństwa w ochronie środowiska oraz o ocenach oddziaływania na środowisko* (consolidated text *Journal of Laws* of 2022, item 1029 as amended).

Waste Act of 14 December 2012 – *Ustawa z dnia 14 grudnia 2012 r. o odpadach* (consolidated text *Journal of Laws* of 2022 item 699, as amended).

Water Law of 20 July 2017 – *Ustawa z dnia 20 lipca 2017 r. – Prawo wodne* (consolidated text *Journal of Laws* of 2022 item 2625, as amended),

Act of 13 July 2023 on the revitalization of the Odra River – *Ustawa z dnia 13 lipca 2023 r. o rewitalizacji rzeki Odry* (*Journal of Laws* of 2023, item 1963).

Regulation of the Minister of Environment of 28 December 2017 on granting the statute to the State Water Holding Polish Waters – *Rozporządzenie Ministra Środowiska z dnia 28 grudnia 2017 r. w sprawie nadania statutu Państwowemu Gospodarstwu Wodnemu Wody Polskie* (*Journal of Laws* of 2017, item 2506).

Regulation of the Minister of Marine Economy and Inland Navigation of 27 August 2019 on the types of investments and actions that require a water-law assessment – *Rozporządzenie Ministra Gospodarki Morskiej i Żeglugi Śródlądowej z dnia 27 sierpnia 2019 r. w sprawie rodzajów inwestycji i działań, które wymagają uzyskania oceny wodnoprawnej* (*Journal of Laws* of 2019, item 1752).

Regulation of the Minister of Marine Economy and Inland Navigation of 12 July 2019 on substances particularly harmful to the aquatic environment and the conditions which must be fulfilled when wastewater is discharged to water or to soil and when rainwater or water from melting snow and ice is channelled away to water or to water engineering structures – *Rozporządzenie Ministra Gospodarki Morskiej i Żeglugi Śródlądowej z dnia 12 lipca 2019 r. w sprawie substancji szczególnie szkodliwych dla środowiska wodnego oraz warunków, jakie należy spełnić przy wprowadzaniu do wód lub do ziemi ścieków, a także przy odprowadzaniu wód opadowych lub roztopowych do wód lub do urządzeń wodnych* (*Journal of Laws* of 2019, item 1311).

Regulation of the Council of Ministers of 10 September 2019 on enterprises with a potential considerable environmental impact – *Rozporządzenie Rady Ministrów z dnia 10 września 2019 r. w sprawie przedsięwzięć mogących znacząco oddziaływać na środowisko* (*Journal of Laws* of 2019, item 1839 as amended).

Regulation of the Minister of Infrastructure of 25 June 2021 on the classification of the ecological status, ecological potential and chemical status, as well as on methods of classification of the status of surface water bodies and environmental quality standards for priority substances – *Rozporządzenie Ministra Infrastruktury z dnia 25 czerwca 2021 r. w sprawie klasyfikacji stanu ekologicznego, potencjału ekologicznego i stanu chemicznego oraz sposobu klasyfikacji stanu jednolitych części wód powierzchniowych, a także środowiskowych norm jakości dla substancji priorytetowych* (*Journal of Laws* of 2021, item 1475).

Regulation of the Minister of Infrastructure of 13 July 2021 on forms and methods of monitoring of surface water bodies and groundwater bodies – *Rozporządzenie Ministra Infrastruktury z dnia 13 lipca 2021 r. w sprawie form i sposobu prowadzenia monitoringu jednolitych części wód powierzchniowych i jednolitych części wód podziemnych* (*Journal of Laws* of 2021, item 1576).

# Index